AF624874

# The 4 Element Synth

To access the online videos visit:
www.halleonard.com/mylibrary
Enter code:

7325-0541-1948-8912

# The 4 Element Synth

## The secrets of subtractive synthesis

Rob Papen

Rowman & Littlefield
Lanham • Boulder • New York • London

Published by Rowman & Littlefield
An imprint of The Rowman & Littlefield Publishing Group, Inc.
4501 Forbes Boulevard, Suite 200, Lanham, Maryland 20706
www.rowman.com

6 Tinworth Street, London SE11 5AL, United Kingdom



**Library of Congress Cataloging-in-Publication Data**
ISBN 978-1-5381-3674-4 (hardcover)
ISBN 978-1-538-13675-1 (paperback)

∞™ The paper used in this publication meets the minimum requirements of American National Standard for Information Sciences—Permanence of Paper for Printed Library Materials, ANSI/NISO Z39.48-1992.

The best way to watch the videos, is in your studio using a computer with a good sound card. When you start to edit sounds on your synthesizer, always keep in mind that audio peaks can occur which could damage your audio system. So be careful, and use low volume settings.

# Welcome

Welcome to this unique masterclass training, in which I want to increase your skills in editing and making new sounds with synthesizers. It doesn't matter if you are a beginner or a pro, this training will give you a solid base and fresh new ideas for working with synthesizers.

I'll focus on subtractive synthesis in this training since this is the most commonly used synthesis for synthesizers. Samplers also fit into this synthesis group.
The method I use is called The 4-Element Synth©.

This method is very powerful because of its clear system and easy accessibility. This is the same well tested training method I use in my masterclass training here in the Netherlands. As well as teaching you a solid foundation, it will help you develop a personal sound. I'll also add many practical tips, tricks and sound examples of how a sound can be made.

It's important to use this book as a guide and to follow it step by step. The 4-Element Synth chart can be found at the very end of this book. In addition to this book and the videos, it is the other main part of this tuition.

After completing this masterclass training, you'll have a very different and more effective approach to your synthesizers and because of that, you'll get a lot more pleasure out of them.

Yours,

Rob Papen

# Contents

# The 4 Element Synth

# The 4 Element Synth

The entire master class is constructed around The 4 Element Synth. In this book, you'll find a printed version of our synth in schematic form. Keep it next to you during the entire session. Let's take a closer look at it now. As you can see, this synth is built using 4 elements. They all have their own color. Green for Element-1, Blue for Element-2, Brown for Element-3 and Yellow for Element-4. This schematic covers all the basics about subtractive synthesis! There are of course, many variations on these concepts, such as synthesizers that have three or four oscillators instead of two oscillators like the one you see in The 4 Element Synth. But the differences are still just variations. To understand the secrets of subtractive synthesis and to learn the art of making and editing presets, The 4 Element Synth has everything we need.

***Important note:***
***The goal of this training is for you to adapt The 4 Element Synth as a way of thinking and an approach towards subtractive synthesis. You'll soon see and come to understand that the chart is indeed a essential key to subtractive synthesis.***

Watching the videos is best inside your music studio using the computer and a good audio card.

More things to learn from the chart. If we look at Element-1 or the Green section, we can see it has darker and lighter parts. These other green part represent different parts within the Element-1. Well, let me tell you, this makes life a lot easier. You can now conclude from studying the chart that an oscillator is made up of two parts. So if you look at an oscillator in one of your synths and see all those knobs and parameters, you'll know by looking back at the chart that it's still only made up of two parts. As you can see, from this point on, the chart is very important in this training.

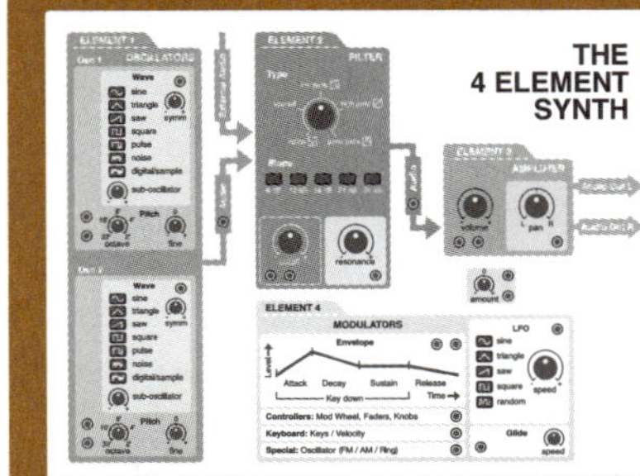

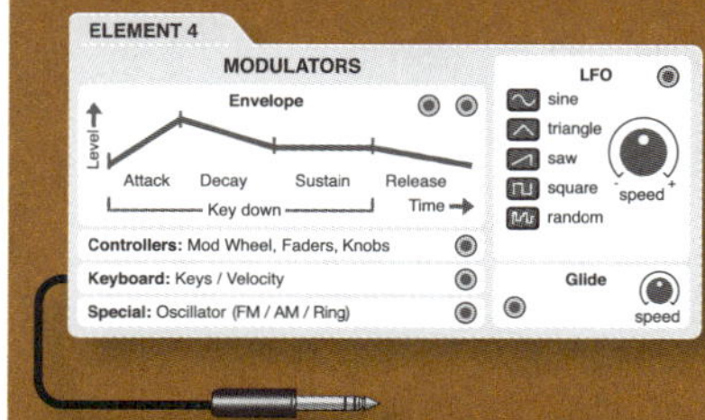


Other observations: We can also see that the chart shows many buttons and knobs. The red dots that you see are inputs. The various parts of Element-4 can be connected to these inputs using virtual cables. If you see 2 inputs inside a part it means that it is used quite frequently and if you see only 1 input it means that it is used less frequently. Of course, in real use, more than 2 inputs can be used. These connections are so important because all of the elements are separated within the synth. You won't hear any sound if you haven't made any connections!

For example, here's a picture of a beautiful Moog Modular from about 1964. If you take this classic synth into your studio, press the 'on' button, connect the audio to your mixer... nothing is going to happen. That's because you first need to connect the separate modules inside the modular synthesizer before you can hear a single note! Back in the day, they really had to study the manual and even get technical help from Moog because there was no such thing as presets ;-)

Moog Modular System 55 (picture Dieter Stork).

Looking back at the chart of The 4 Element Synth... you don't see any connections except for audio that is routed out of Element-1 then goes into Element-2 and then into Element-3. Just like our 'Old School Moog Modular', we need to connect things to each other to make anything happen in The 4 Element Synth. For learning and understanding the principles of subtractive synthesis these connections are very important. So let's first dive into the topic Element-4 in the next chapter entitled 'Before you go on'.

Again, as I said before, keep 'The 4 Element' chart next to you during the entire training for quick reference.

***Introduction: 'Introduction 4 Element Synth.***

***Introduction: 'Brief introduction subtractive synthesis'.***

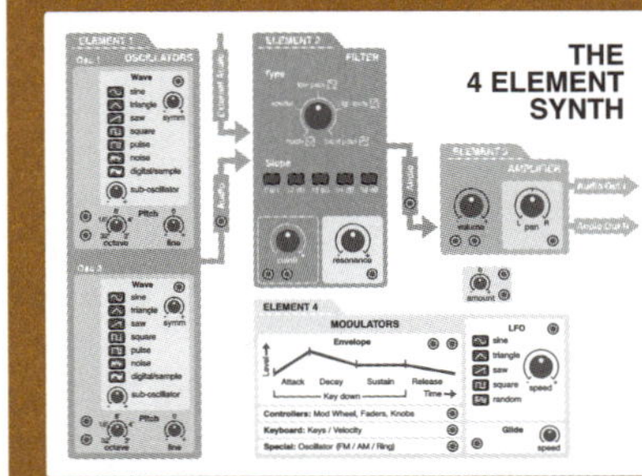

# Before you move on

# Before you move on

Before we explain Element-1 and all its features, it's important to look at Element-4 first. If we go from left to right on The 4 Element Synth chart, we go from Element-1 (green) to Element-2 (blue) and then to Element-3 (brown). This is what we call the audio path of the synthesizer.

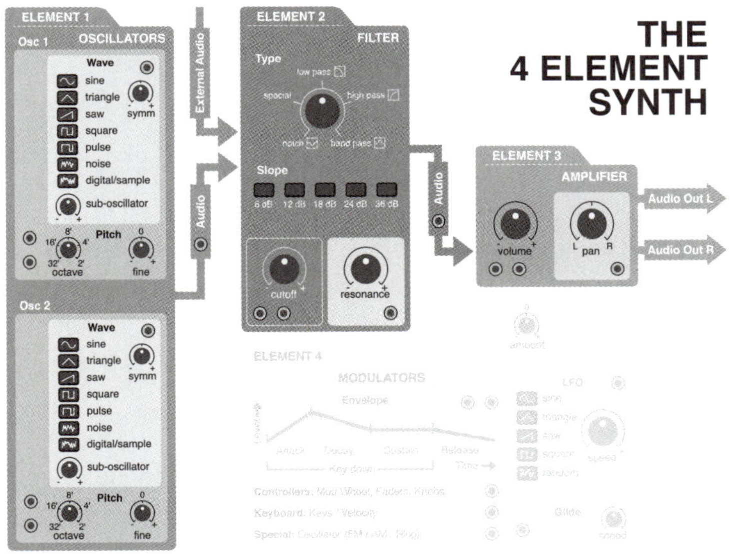


These are the parts you actually hear. Everything that comes out of the audio output of Element-3 travels in this order: Element-1 ⇨ Element-2 ⇨ Element-3. You won't find an Element-4 (yellow) in this audio chain because this Element-4 is the central controller of the synthesizer. It controls all the other elements.

You can compare it to your brain. Without your brain you couldn't move your legs, arms, speak or even read this book...

So Element-4 controls all the other elements. That's why you need to read this chapter and look at the Video section *'Before you go on'* because it's important to understand what the basic function of Element-4 is!

# THE 4 ELEMENT SYNTH

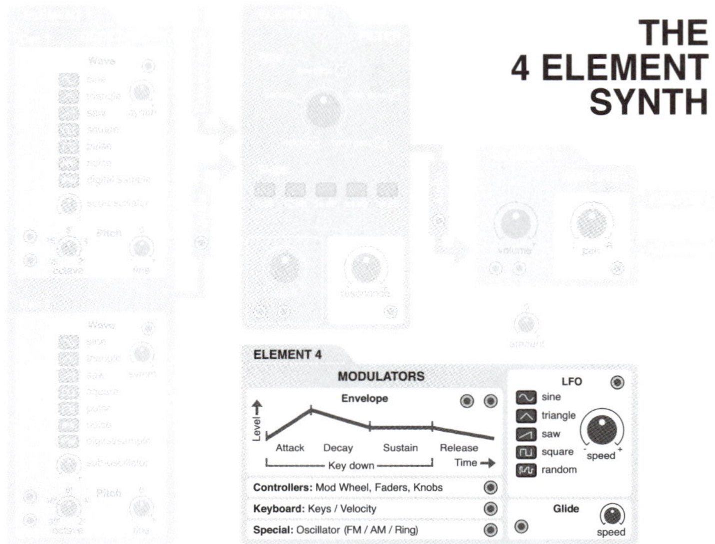


In general, we talk about modulators as seen in the Element-4 image. You may have heard the term modulation before. For sure you've heard the term modulation wheel. It's the wheel next to the pitch bend that changes the sound when you move it. Generally, modulators all do the same thing. They change the parameters that you assign them to change.

As you can see, the whole section of Element-4 is full of modulators and they are responsible for all types of dynamic actions as well as controlling the other elements. It can even control parts inside itself.
Now watch the video '*Before you go on*'. It explains the things you just read in more detail. After you have watched this, you'll understand the main function for the Element-4.

***The key to Element-4 is:***
***It makes the other elements work and adds dynamic changes.***

***Introduction: 'Before you move on'.***

# Element 1

# Element-1 / Oscillators

*Introduction*

**Introduction**
The basis of each sound starts at Element-1. It's the green part on the 4 Element chart. Take the printed chart and look at Element-1. If you look closely at the chart you'll see that the Element-1 is named oscillators. This term and its variations are used on every subtractive synthesizer.

In the Element-1 you can see that there are two independent oscillators, oscillator 1 and oscillator 2. There are synthesizers on the market with more than two oscillators of course, but for explaining subtractive synthesis, two oscillators are more than enough. The key word for Element-1 (green) is oscillator!

If we take a closer look at one of the oscillators, we see that it's built around two components called pitch and wave. Both components are separated by using different colors of green.

*Remember this:*
- ***Element-1 is the basis for each sound and has oscillators.***
- ***Element-1 has two independent oscillators, oscillator 1 and oscillator 2.***
- ***Each oscillator has two main components called pitch and wave.***

*Element-1: 'Introduction Element-1'.*

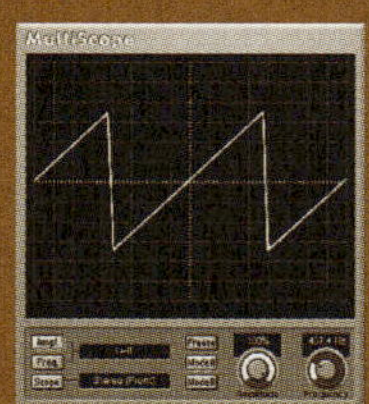

*Waveforms*

## Waveforms part 1

In the paragraph 'Introduction' of Element-1 we've learned that Element-1 is the basis for each sound and that the main part is called an oscillator. These oscillators use waveforms for generating their basic sound. We have two oscillators in our Element-1 but for explaining the component waveform we'll use only one oscillator as an example.

Waveforms used in subtractive synthesis are displayed as names and/or symbols. If we look at Element-1 we can see them and I am sure you have seen these symbols and names before. We have the sine waveform, triangle waveform, saw waveform, square waveform, pulse waveform, noise waveform, digital or sample waveform and the sub-oscillator.

The waveform symbol has a certain look as you can see in the MiniMoog picture and if you connected the synthesizer to an electronic device called an oscilloscope the wave will be visible and will look identical to its symbol! The sine symbol looks just like a sine wave and the saw wave looks like the saw symbol.

What more can we say about waveforms? Waveforms contain the basic harmonic components of each sound. The word harmonic or harmonics is maybe new to you. Let's go into greater detail about this, because it's really essential to know.

WATCH ONLINE

***Element-1: 'Waveforms part 1'.***

## Harmonics

*Harmonics*

As you can see this is not a synthesizer but a guitar. Why? Well, to understand harmonics it's a good idea to look at the acoustic world. Many of the physical laws of sound and physical events that exist in the acoustic world are the same in subtractive synthesis as well. Even more than you think!

When you strike a string you can hear that it's a guitar and also that you're playing a note, for instance an E. That you recognize its sound as a guitar and the fact that it's sounding the note E is determined by two facts:

1. Fundamental

The fundamental allows you to determine that it's the note E. Every sound that has a definable pitch has a fundamental. This fundamental is the basic frequency of the tone. Our example note E has a fundamental frequency of 329.6 Hz. The fundamental frequency is present in the tone as a sine waveform of 329.6 Hz. This image shows the fundamental.

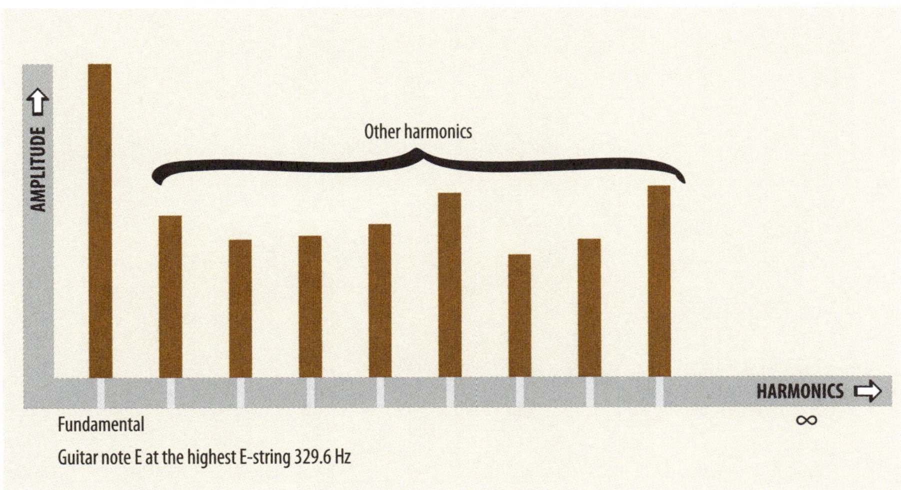


Figure 1.01

Fundamental.

2. Harmonics

Above the fundamental frequency there are other pure sine waves that together form the character of an instrument or sound. This is what we call harmonics. These sine waves are at different frequencies and volumes above the fundamental. The image below shows the fundamental with harmonics.

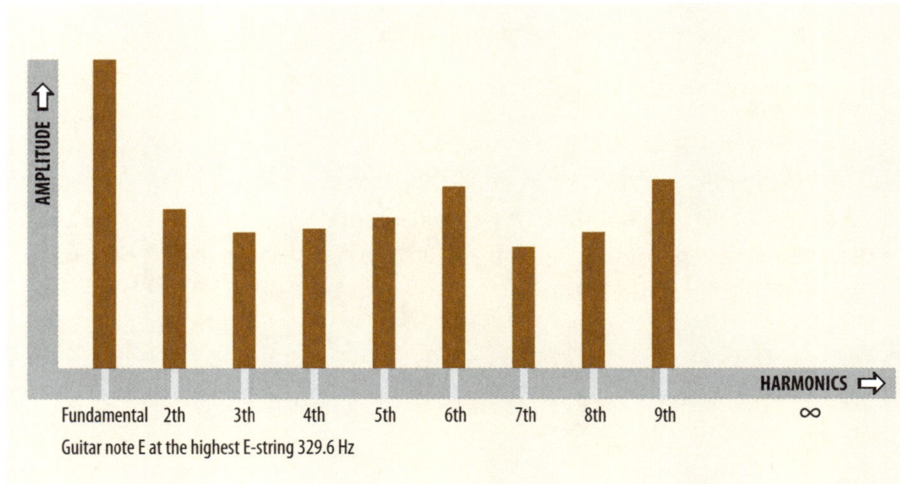


Figure 1.02

Harmonics.

In such an acoustic instrument as the guitar there are countless harmonics and each has its own volume curve in time. For example, this means that a harmonic which is at 1.000 Hz can last for 2 seconds with a certain volume curve and that another harmonic that is at 18.000 Hz can last for only 1 millisecond.

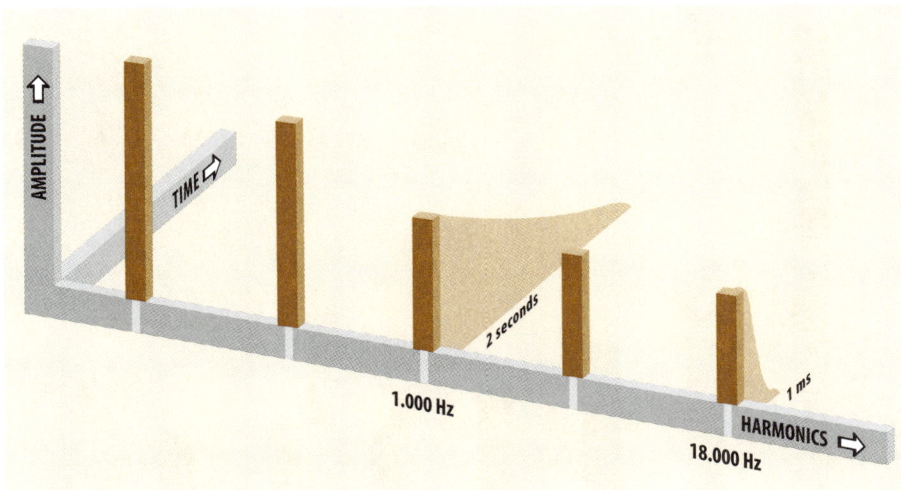


Figure 1.03

Each harmonic has its own volume curve in time.

Harmonics in an acoustic sound move in a very complex manner in time. If we go back to a subtractive synthesizer, we find that these harmonics are also inside most waveforms. But they are static in volume, instead of being dynamic. So every harmonic is always at a certain volume level.
Shown below are the harmonics of a saw waveform. As you can see the harmonics form a kind of logical structure. To understand sound and synthesizers it is important that you know what the fundamental and harmonics are.

Figure 1.04

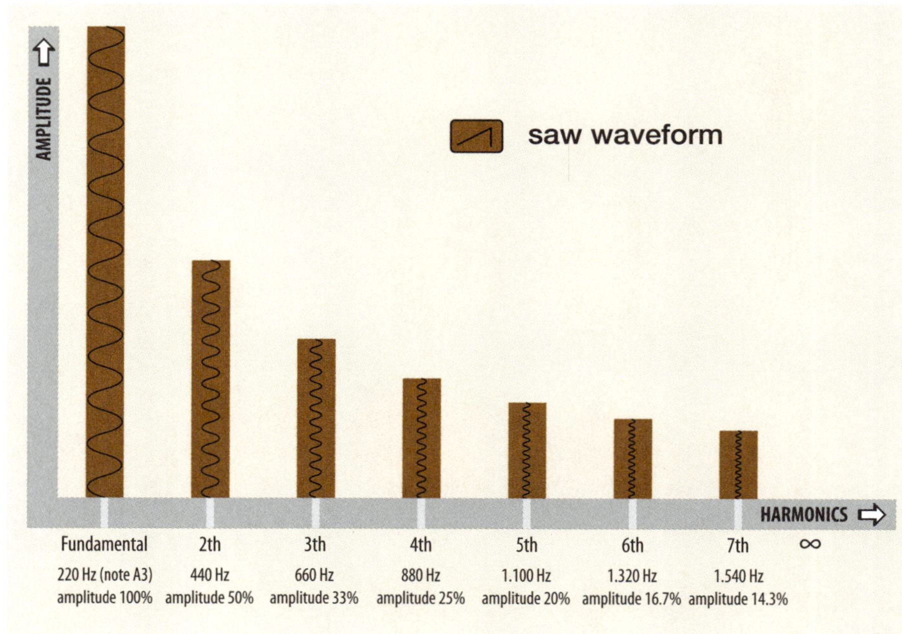


The harmonics of a saw waveform.

At a later point, in several places in this training you will see that these harmonics and the fundamental determine the sound and have a lot of effect on the sound, even in places where you may not expect it.

WATCH ONLINE ***Element-1: 'Harmonics'.***

## Waveforms part 2

*Waveforms part 2*

Oscillators use waveforms for generating their basic sound and the character is formed by the harmonics inside each waveform. As we discussed in the chapter *'Waveforms part 1'* the waveforms used in subtractive synthesis are displayed as names and/or symbols. If we look again at the Element-1 we can see the sine waveform, triangle waveform, saw waveform, square waveform, pulse waveform, noise waveform, digital or sample waveform and the sub-oscillator.

The question is, of course, how do we use them? Below you see a list of The 4 Element Synth waveforms. For each waveform, we've provided you with a reference instrument, some examples of instruments that use that waveform as well as a character description.

The reference instruments or sounds have an important purpose in this training. If you want to build your own sounds, then you need to know what waveform type you need to start with. Remember that the foundation of a synth sound starts here in the Element-1 section. So you'll need to pick the right waveform from the beginning to get the results you want!

| Waveform | Instrument/reference | Other Instrument examples | Character |
|---|---|---|---|
| Sinus | Whistle, test tone | Singing saw, organ, sub-bass | Pure tone |
| Triangle | Recorder | Church organ, flutes | Woody sound, mellow |
| Saw | Trumpet | Synth basses, synth brass | Brassy |
| Square | Clarinet | DnB bass, clarinet | Hollow sounding |
| Pulse | Clarinet ⇨ oboe | Clavinet, electric bass, guitar | Hollow ⇨ nasally sounding |
| Pulse modulated by LFO (PWM) | Pad and chorus-like | Synth sounds, hardstyle leads | Moving and chorus-like |
| White Noise | Wind | Sound FX | High sounding noise |
| Pink Noise | Explosion | Thunder, big drum | Deep sounding noise |
| Digital | Bell | Digital wavetable synthesizers | Very different character |
| Samples | Acoustic instruments | Piano, vocal sounds | Acoustic and/or different |

*Element-1: 'Waveforms part 2'.*

### Sinus waveform

The special character of this waveform is that is has no harmonics, only the fundamental. So what you hear is the pure tone or pitch. This waveform is in other words totally empty, a blank slate. No harmonics... only a pure tone. This means that this waveform doesn't have much to offer when you add filters (which we'll explain in the chapter entitled Element-2). They are a waste of time to use because there are no upper harmonics which can be filtered. Nevertheless, this pure fundamental tone can be very musical and very useful for certain sounds.

Figure 1.05

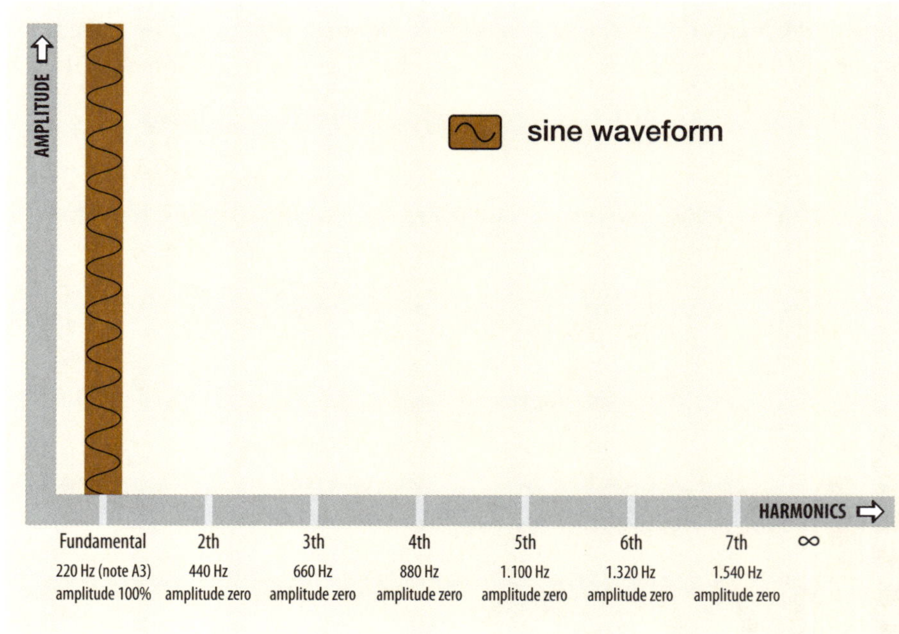


Sine waveform.

| Wavefor | Instrument/reference | Other Instrument examples | Character |
|---|---|---|---|
| Sinus | Whistle, test tone | Singing saw, organ, sub-bass | Pure tone |

WATCH ONLINE *Element 1: 'Waveform examples / Sinus'.*

### Triangle waveform

The sound of this waveform is similar to the sinus, but it sounds more silkier and mellower. In fact, it has a few harmonics that add this character. But, if you were to play this waveform by itself, it's easy to confuse it with a sinus wave.

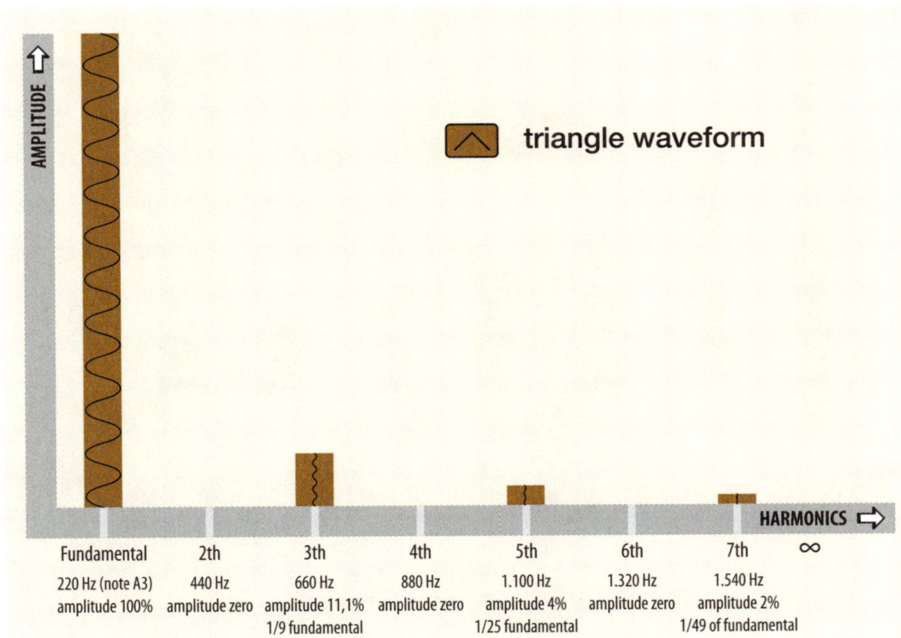


Figure 1.06

Triangle waveform.

| Waveform | Instrument/reference | Other Instrument examples | Character |
|---|---|---|---|
| Triangle | Recorder | Church organ, flutes | Woody sound, mellow |

*Element-1: 'Waveform examples / Triangle'.*

## Saw waveform

This waveform is very versatile and the reason for that is its harmonics. It contains all the possible harmonics -both even and odd ones-, starting from the fundamental on up.

Figure 1.07

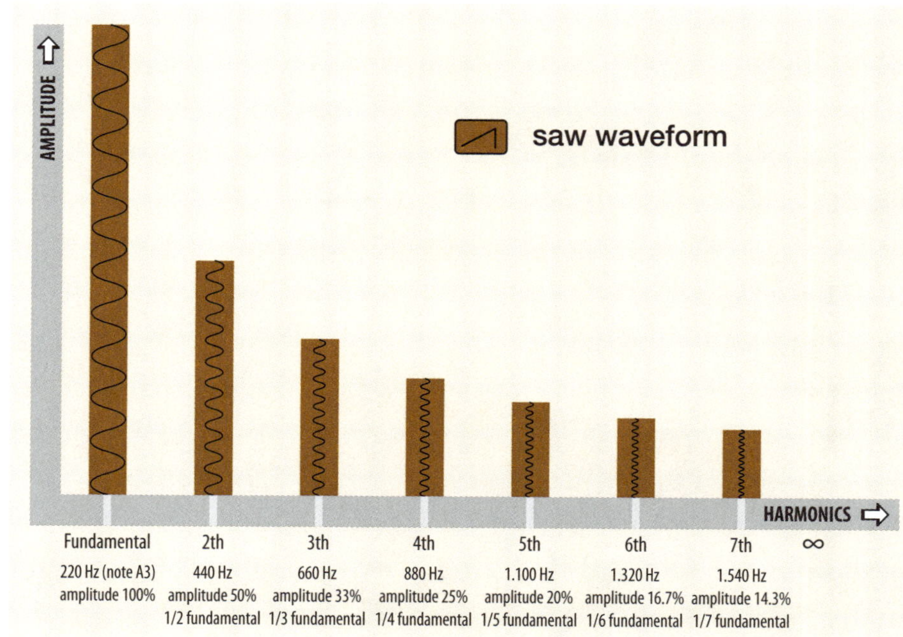


Saw waveform.

| Waveform | Instrument/reference | Other Instrument examples | Character |
|---|---|---|---|
| Saw | Trumpet | Synth basses, synth brass | Brassy |

***Element-1: 'Waveform examples / Saw'.***

### Square/pulse 50% waveform

The square wave is something special. What makes it special is that in many synths you can change the width of the square wave. This changes the square wave into a pulse wave. Basically if you compare a pulse waveform with 50% symmetry and a square wave, both are identical! Note that a square wave is the same as a pulse wave that has a symmetry of 50%. See the image below.

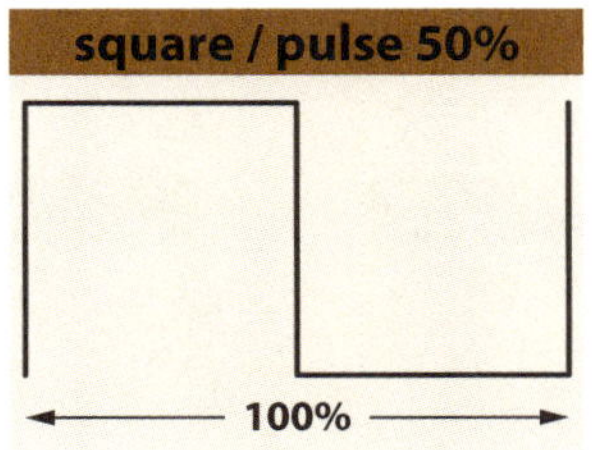


Figure 1.08

*Element-1: 'Waveform examples / Square and Pulse'.*

### Symmetry

Of course you are thinking to yourself....what does symmetry mean? Look at the picture with the square wave and the pulse wave below.

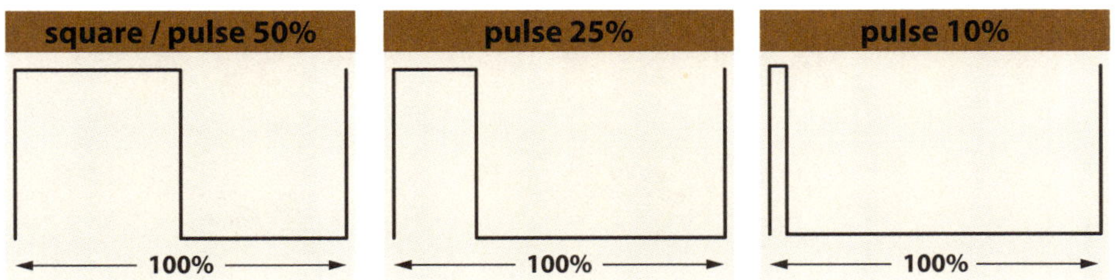


Figure 1.09

A waveform has a certain length. The picture shows the waveform's entire length. Another word for this is a cycle. When we're talking about symmetry on a synth, we're talking about changing the midpoint of the wavelength of a waveform. This can be done to all the waveforms.

But a change in the symmetry is most effective on square or pulse waveforms. This is because you can hear a big difference in the way it sounds when you change the symmetry of the square and pulse waveforms. See below the picture of the pulse waveform and saw waveform. Both are also displayed with a different symmetry. This illustrates that it's also possible to change the symmetry of a saw waveform. But the audio result of the change is not very effective. It makes the saw sound rather thin.

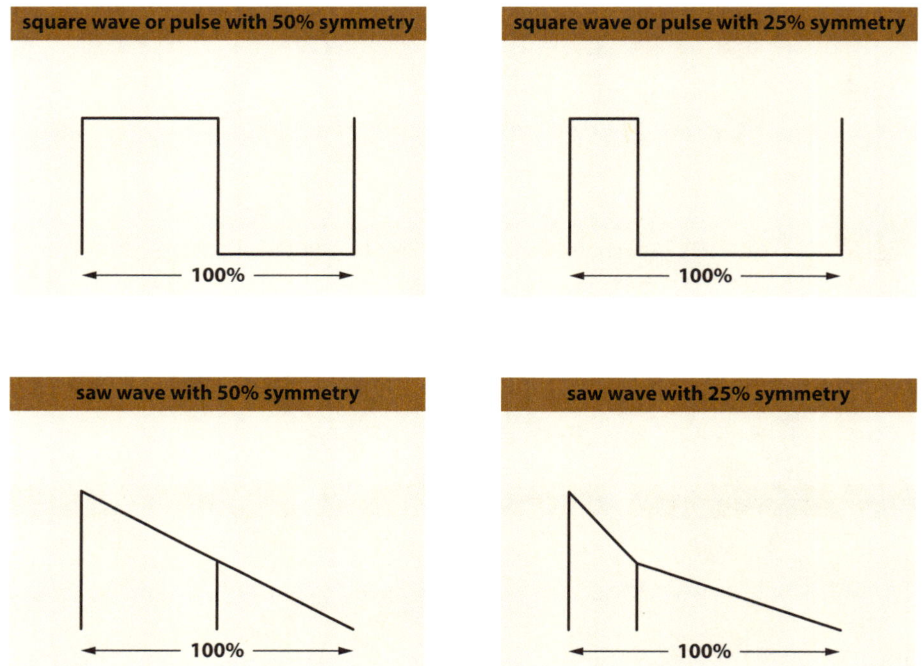


Figure 1.10

If we look at the classic analog synths, we see only the name pulse width. Because of the fact that it is most effective on the pulse or square wave, they simply left out this function for the other waveforms. They also called the name of the function pulse width instead of calling it symmetry.

This is logical after all, because if you change the symmetry of the pulse waveform, you're changing the width of the pulse. Look at the image again. Take note that on older synths you're going to see pulse width, but basically it is the same as symmetry... and it's only used on the square or pulse waveforms.

***So remember:***

- ***the symmetry control changes the midpoint of a waveform.***
- ***symmetry is the same as pulse width, but is only used on a pulse or square waveform.***

***Element-1: 'Waveform examples / Symmetry modulation'.***

With today's synths it's possible to change the symmetry on other waveforms. On the Video we'll show you a few examples of what you can do with symmetry on other waveforms. The results are not always good, such as on a saw wave but using it on digital waves or sinus waveforms it can still be interesting.

## Now back to the square waveform

In the image below you can see the harmonic content. As you can see it contains only odd harmonics and they are built up in volumes according to a certain order or rule. For instance, the volume of the 3rd harmonic is 1/3rd of the total volume and the 5th harmonic is 1/5th of the total volume. Once you change the symmetry or pulse width, the composition of the waveform changes and other harmonics are added.

Figure 1.11

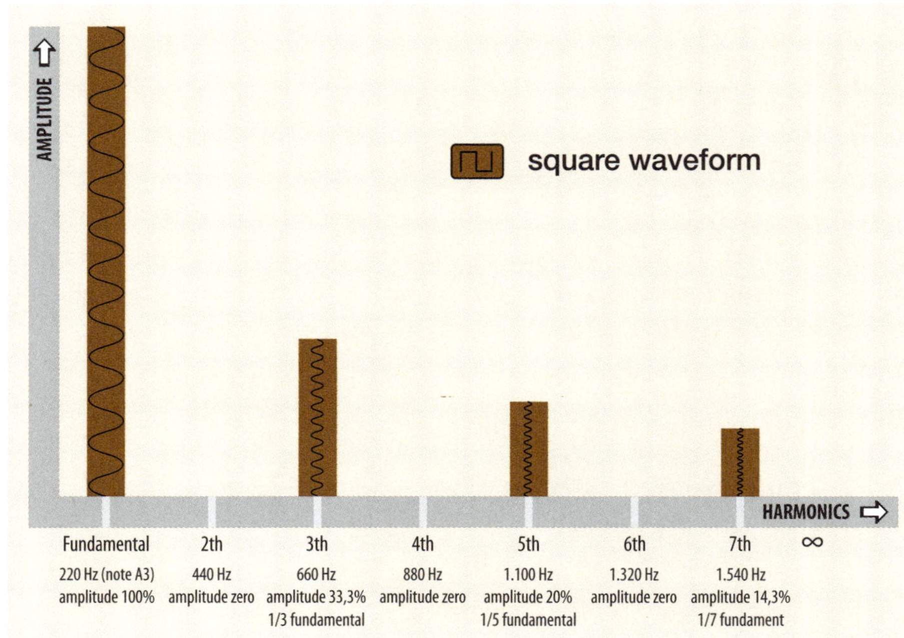


Square waveform.

| Waveform | Instrument/reference | Other Instrument examples | Character |
|---|---|---|---|
| Square/Pulse 50% | Clarinet | DnB bass, Clarinet | Hollow sounding |

***Element-1: 'Waveform examples / Square'.***

### Pulse waveform

Once you start changing the pulse width, the composition of the waveform changes. Depending on the pulse width value you choose, certain harmonics are left out from the series of harmonics, and next to that the volume of the remaining harmonics changes.

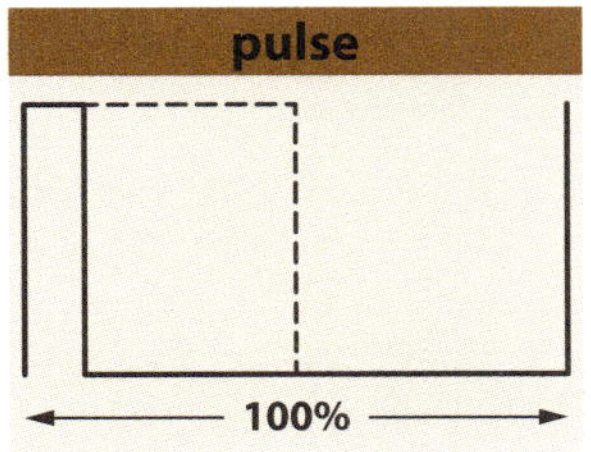


Figure 1.12

You notice that the square waveform or pulse waveform if you change the width of it is very versatile. Of course you can change the width also to 75% instead of 25%. But sound wise this makes no difference.

This pulse width part that you can change, can be done automatically by using one of the Element-4 parts. This is something we'll discuss in the Element-4 and also on the Video along with some sound examples. So please continue now on the Video to hear the square/pulse waveforms and what you can do with them.

| Waveform | Instrument/reference | Other Instrument examples | Character |
|---|---|---|---|
| Pulse | Clarinet ⇨ oboe | Clavinet, electric bass, guitar | Hollow ⇨ nasally sounding |
| Pulse modulated by LFO (PWM) | Pad and chorus-like | Synth sounds, hardstyle leads | Moving and chorus-like |

*Element-1: 'Waveform examples / Pulse'.*

## Noise (white/pink)

Noise is the perfect word to describe this waveform. It generates a lot of noise. In many synthesizers we have the option to choose between white noise and pink noise. You can hear that pink sounds far more darker and heavier than white.

Noise basically contains all frequencies, the difference is how the total energy of the sound is distributed over these frequencies. White noise has equal energy throughout the entire frequency spectrum. Therefore, the white noise sounds lighter to our ears: The higher the frequencies, the relatively louder they are.

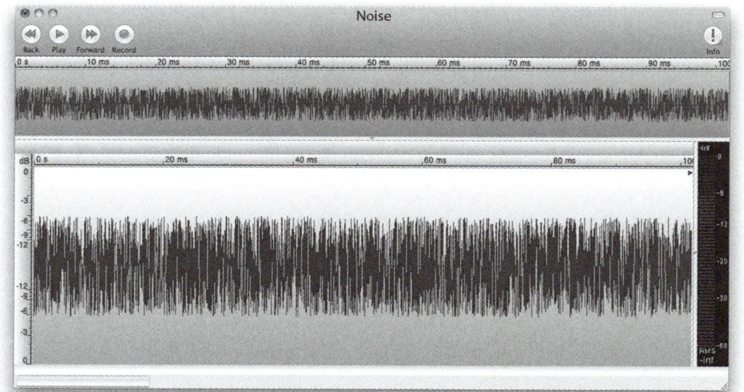


Figure 1.13

Noise.

| Waveform | Instrument/reference | Other Instrument examples | Character |
|---|---|---|---|
| White Noise | Wind | Sound FX | High sounding noise |
| Pink Noise | Explosion | Thunder, big drum | Deep sounding noise |

With pink noise every octave has equal energy throughout the entire frequency spectrum, which makes it sound heavier. As a general rule, to our ears, low frequencies sound louder.

WATCH ONLINE ***Element-1: 'Waveform examples / Noise'.***

**Digital/sample waveforms**
With the saw waveform or triangle waveform the construction of the harmonics follows a certain compositional pattern. With digital waveforms this composition can be free or can be made by analyzing samples. In this case software analyzes one cycle of the audio and reproduces it as a waveform. Of course this waveform is static in sound, because it is only a single waveform. Still, it can be very useful and great for creating new and exciting presets. Recognizing these waveforms is very easy, because they don't fit into the references that we have learned to distinguish the other waveforms with such as sinus, triangle, saw, square and pulse. They often have a digital character or contain the character of an analyzed sample.

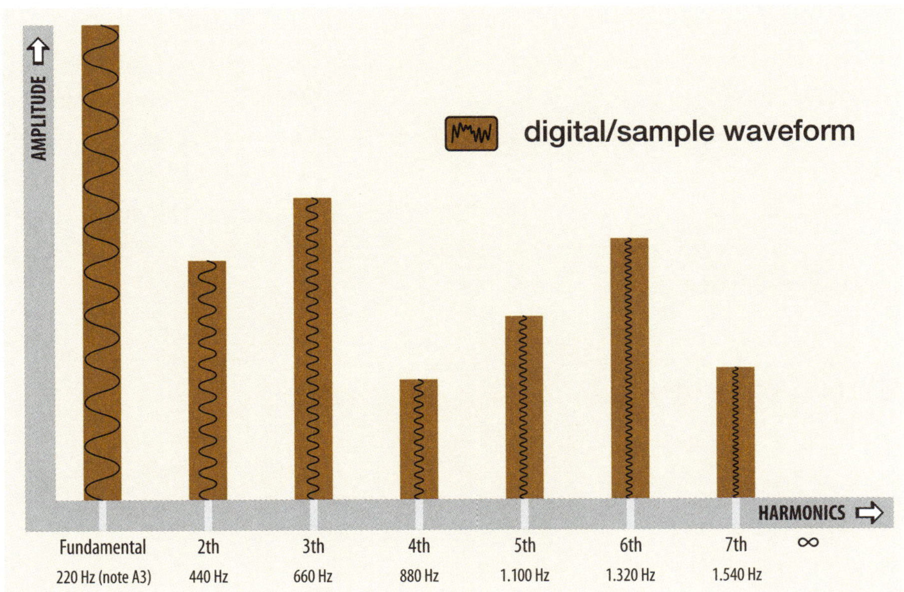


Figure 1.14

Random composition of a digital waveform.

| Waveform | Instrument/reference | Other Instrument examples | Character |
|---|---|---|---|
| Digital | Bell | Digital wavetable synthesizers | Very different character |
| Samples | Acoustic instruments | Piano, vocal sounds | Acoustic and/or different |

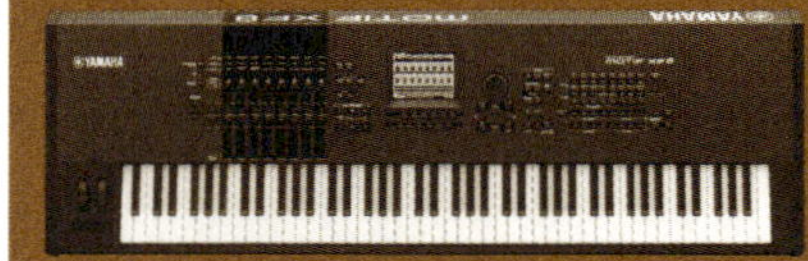

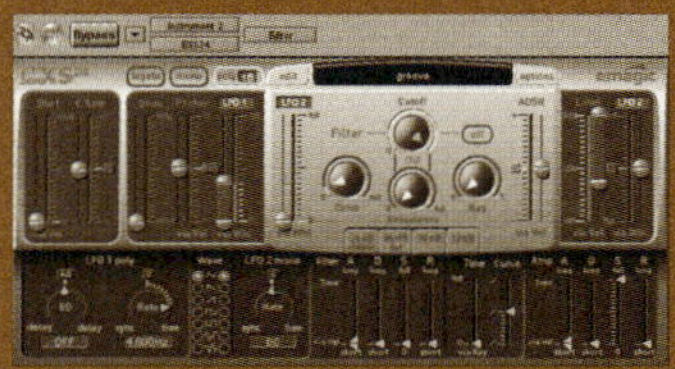

Most workstations and romplers have sampled oscillator waveforms. A ROMpler is a type of synthesizer that contains a huge library of samples within the synth that are stored into a ROM memory that can't be changed. Because these samples are placed into a ROM memory instead of a RAM memory, we call this type of synth ROMpler

You usually don't have the option to sample sounds for yourself. You can only use the samples already in the synth. Most ROMplers can handle bread and butter sounds of all kinds. Of course, a sampler also uses sampled waveforms, but the difference is that you can make your own samples or you can load them in from a hard disk or CD-ROM/DVD.

In terms of actual synthesis, almost all samplers, workstations and romplers use subtractive synthesis. So there is no need to study another type of synthesis! The only difference between these instruments and a regular subtractive synthesizer is the fact that the waveforms such as saw or triangle waveforms are not being generated by the synthesizer itself, but instead are samples or collections of samples that are being used as waveforms. Of course..., no one said you can't sample saw and triangle waveforms.

WATCH ONLINE

***Element-1: 'Waveform examples / Digital and Sample.***

**Sub-oscillator**

A sub-oscillator is not an independent oscillator and that's why it's called sub osc. Some might think that it generates some kind of sub low frequencies or that it increases the bass-amount of a sound.

What the sub oscillator really does is divide the oscillator so that a waveform one octave lower is generated. Most of the time it generates a square waveform. So, if I select a saw waveform and open the sub-oscillator, a square waveform one octave below the saw waveform is generated.

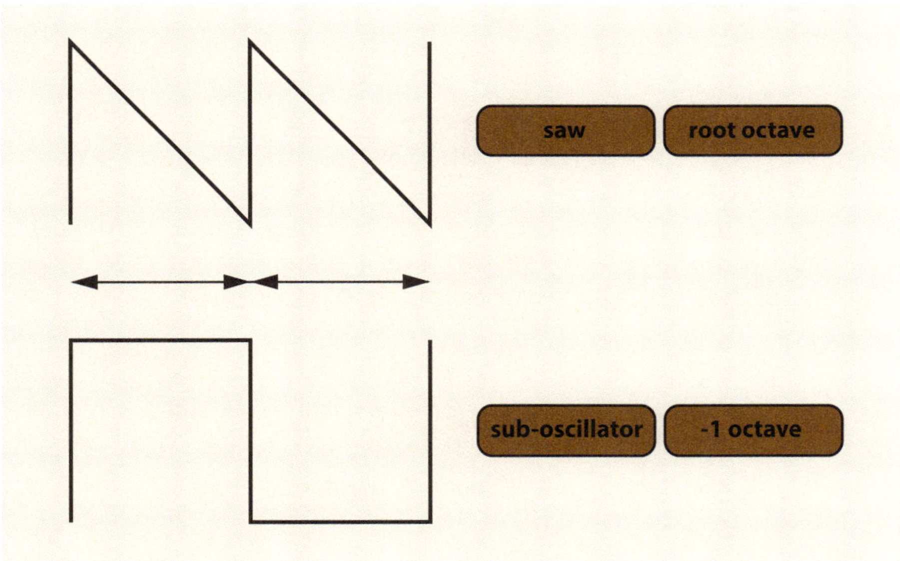


Figure 1,15

Sub-oscillator.

*Element-1: 'Sub-oscillator'.*

*Waveform modulation*

## Waveform modulation

What happens with waveform modulation is that you connect a part from Element-4 and it starts changing the symmetry of the waveform. So without connecting Element-4, waveform modulation is not possible at all.

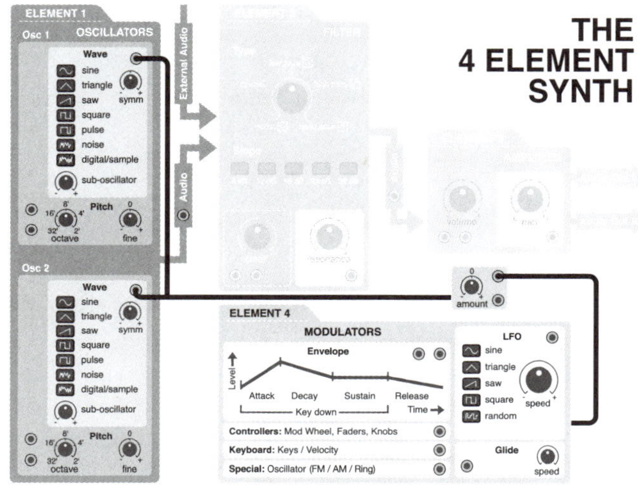


Figure 1.16

The key for Element-4 is: "It makes the other elements work and adds dynamic changes". At older synths you see only pulse width, but basically it's the same as symmetry... it's addressed only at the square or pulse waveform.

*So remember:*

*• Waveform modulation means that a part from Element-4 changes the symmetry of a waveform.*

*• Symmetry is the same as pulse width, but used on a pulse or square waveform.*

*Element-1: 'Waveform modulation'.*

There are also other types of waveform modulations. These types of modulations became possible with the entrance of the digital synthesizers. One of the very first ones introducing phase-distortion was the Casio CZ range in 1984. It used the phase-distortion as a complete synthesis.

What happens with phase-distortion is that the phase angle of the waveform is bent. With this phase angle changed, other harmonics are added a new waveform is created. Wave-shaping is close to phase-distortion but it alters directly the shape of the wave instead of just the phase. A synthesizer which is able to do phase distortion is for instance the virtual synthesizer Rob Papen BLUE.

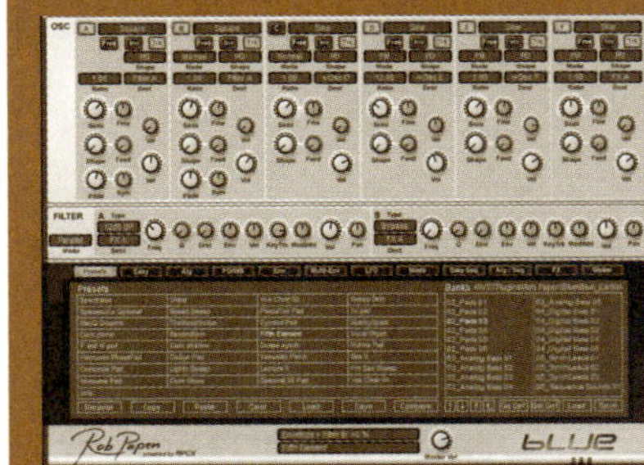

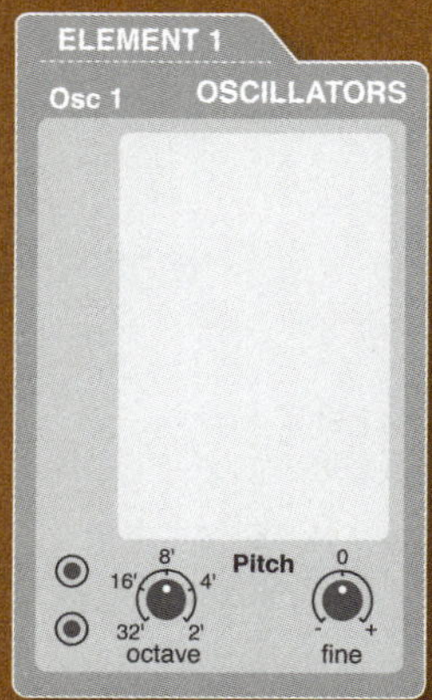

*Oscillator pitch*

## Oscillator pitch

In the chapter introduction of Element-1 we learned that the Element-1 is at the core of every sound and that the main building block of a sound is called an oscillator. These oscillators use waveforms for generating their basic sound. We have two oscillators in our Element-1 and that is enough to explain the use of oscillators. If we take a closer look at one of the oscillators, we see that each one is built around two parts called pitch and wave. In the section entitled waveforms, we learned about the waveform part of an oscillator. If we look at the chart again and examine one of the two oscillators we can see that the pitch part has two controls and also inputs.

The two control knobs tune the oscillator. One is the octave knob and the other one is the fine tune knob. We also have these inputs inside the pitch part. Having just two inputs is mainly symbolic, because in practice we can have many more inputs than just two. For our discussion, though, the two inputs on the 4 Element Synth chart signifies that they are used often and that many parts of Element-4 are and can be connected to it.

An oscillator doesn't know what pitch to play unless you connect the keyboard to it. Most of the time, this connection is a standard one, but with many synthesizers it can be switched off as well.

For instance, on a Virus-b, if you press the edit button of the oscillator section you see a screen called key follow. This word describes this function very well. It dictates how the oscillator follows the keyboard. Why do we have this function? Well for one thing, at times, it can be really useful to be able to switch the keyboard on/off. Shutting the keyboard control off can be handy in creating percussive types of sounds, for instance bass drums.

***Element-1: 'Oscillator pitch / In general'.***

### Oscillator pitch octaves detuning

Usually, on most oscillators you can choose the octave setting. For instance, on the MiniMoog you can see that range stands for the octave setting of the oscillator. This tunes the oscillator and if I use two oscillators I can either decide to keep them at the same range or detune them in octave steps.

The type of sound you are planning to make will determine how you set these. For example, if you're creating a bass sound, you'll naturally want to use lower octave settings and of course for creating lead sounds you'll want to use higher octave settings. By using more than one oscillator the octave function does more than just change octaves, it also changes the sound, the octave turns into a sound-shaping tool as well.

On many of today's synthesizers, the octave knob is named semi because you can set the values between the octaves. These values are usually incremented in half steps, which gives you 12 semitones per octave.

WATCH ONLINE *Element-1: 'Oscillator pitch / Octaves detuning'.*

### Oscillator pitch fine detuning

A more delicate change to a sound is created by using the detune function of the oscillators. If you're using only one oscillator, then of course, the detune function is useless. You're only hearing the one oscillator and by detuning it, the result is that you're playing out of tune. You need two oscillators to effectively detune one against the other.

So when your goal is to create musical sounds, it's best not to use fine detuning if you're only going to use one oscillator. You really need two or more oscillators to effectively detune one against the other. Most of today's synthesizers offer a fine detuning range of 100 cents. That is the range of one semitone. The fine detuning parameter can have positive or negative values.

Once I start detuning oscillator 2..., you can hear that the character of the sound changes quite a bit. If I use only a little detuning..., it sounds like a classic Synth Lead sound, but if I detune it a bit more deeply..., it suddenly becomes a Dance lead sound. In between these two settings there are countless variations possible each with their own sound character. If you have an octave difference between the two oscillators, the effect of detuning is different.

The fact of the matter is, you don't want to detune too much. If you detune it too deeply it will make the whole sound out of tune. That's not to say that a bit of deeper detuning can't be used to create a nice contemporary lead sound.

WATCH ONLINE *Element-1 'Oscillator pitch / Fine detuning'.*

**Oscillator pitch fine detuning with more than two oscillators**

So far, we've only used two oscillators to describe the huge influence detuning can have on a sound. But what if a synth has three oscillators or even more? Well, then you can do some nice tricks! When three oscillators are being used the trick is to not detune one oscillator, but rather, leave one on pitch. The other two oscillators are detuned.... one has a positive detuning and the other one has a negative detuning. By adding the third oscillator the sound becomes fuller. This is very useful for certain types of sounds, such as pad sounds.

So, in general, if you use 3 oscillators, it's best to detune one positive and the other one negative and one not detuned at all. If a synthesizer has more than 3 oscillators, you may use them all of course. But keep in mind that more is not always better. It depends on the type of sound you want to make and the musical context for that sound.

WATCH ONLINE *Element-1: 'Oscillator pitch / More than two oscillators'.*

## Pitch modulation

*Pitch modulation*

Introduction

In addition to the octave knob and the fine tune knob we also have inputs inside the pitch part of an oscillator. This means that you can connect parts from the Element-4 into the pitch part. One connection that you've already seen in the pitch chapter and the Before you move on chapter, was the keyboard. That's very much needed if you want to play melodies with a synthesizer. As you may already know, the keyboard is also an Element-4 part which controls the pitch and therefore utilizes pitch modulation. Remember also that other parts from the Element-4 can also be connected to the pitch input of an oscillator. The function of the Element-4 is: "It makes the other elements work and adds dynamic changes". The keyboard connection is a basic pitch connection. We explain this further in the next chapter. Vibrato is another widely used main pitch modulation which we'll discuss next.

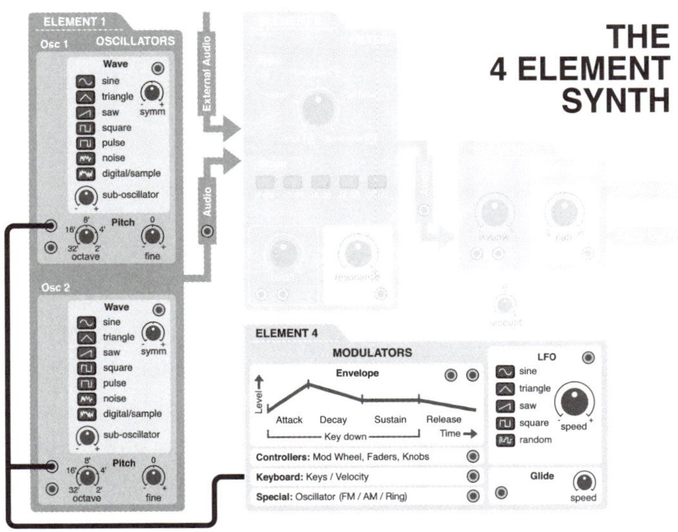

Figure 1.17

### Vibrato (pitch modulation by an LFO)

The LFO is the most commonly used device for creating general pitch modulation. On some synthesizers pitch modulation is called frequency modulation. You'll recall that pitch and frequency are the same. The LFO, sometimes also called modulation generator, is a part from the Element-4 that generates continuous changes. This is perfect for making a classic vibrato effect.

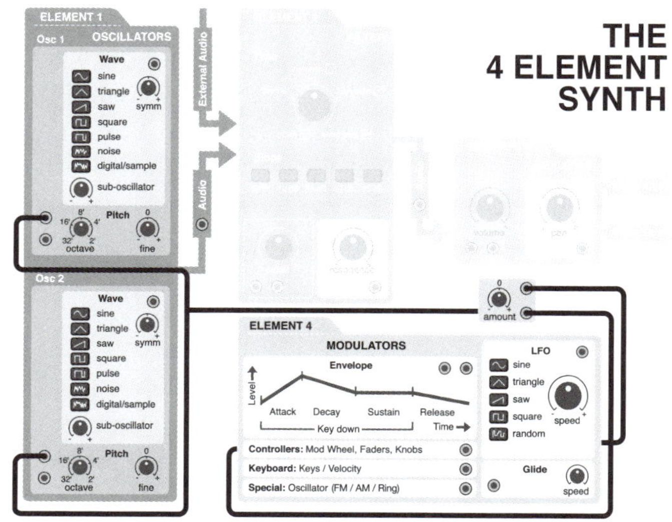


Figure 1.18

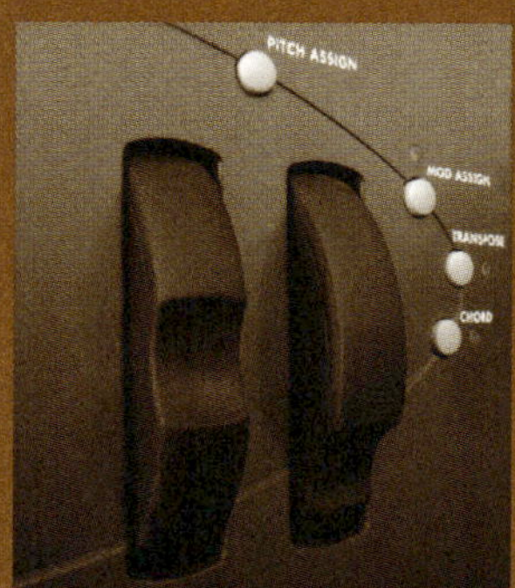


Many synths have a modulation wheel and that is often used to control the pitch modulation by an LFO. The LFO is connected to the amount knob inside the 4ES and then goes from there to both oscillators. Also connected is the modulation wheel and this is connected to the amount knob. So what's really happening is that the modulation wheel controls the amount of pitch modulation. In addition to the modulation wheel, you can also use other controllers or parts from the Element-4 to control pitch modulation. For example, after-touch or even an envelope can be used to control pitch modulation. Check for more pitch modulation info chapter Element-4.

***Element-1: 'Oscillator pitch modulation'.***

## Oscillator sync

If we look at the 4 Element synth chart and the Element-1 section, you can see a hand. This hand symbolizes the sync function between oscillators. At the 4 Element synth oscillator two syncs with oscillator one. At other synthesizers this could be different, so in that case oscillator one could sync to oscillator two. But for understanding what sync does, it makes no difference.

This sync has nothing to do with midi-sync or midi-clock. It is about synchronizing oscillators. Let me try to explain it. Synchronizing oscillators means oscillator two grabs oscillator one and therefore doesn't have a life of its own anymore. It gets connected internally to oscillator one when it comes to frequency, that's why we used the hand as symbol for explaining sync. Every time the master oscillator's cycle repeats, the SYNCed oscillator is retriggered. Regardless of its phase at that particular moment, it starts at the beginning of its cycle.

The oscillator that is synchronized to another oscillator is often called the 'slave oscillator'. The oscillator that is controlling the slave is called the 'master oscillator'. So in case of oscillator 2 synchronizing to oscillator 1, this means that oscillator 2 is the slave and oscillator 1 is the master.

If you've synchronized an oscillator to a master and change its pitch parameter, it will not change in pitch but instead in sound. See the picture below for two examples of synced oscillators and watch the output of the 'slave' oscillator.

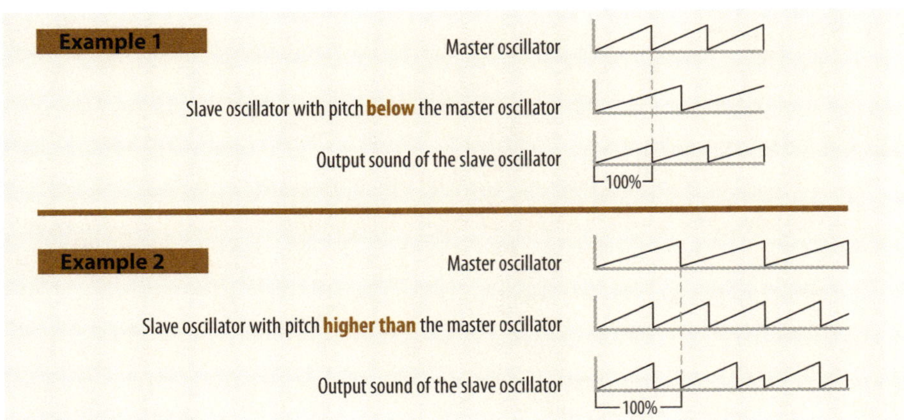


Figure 1.19

Synced oscillator.

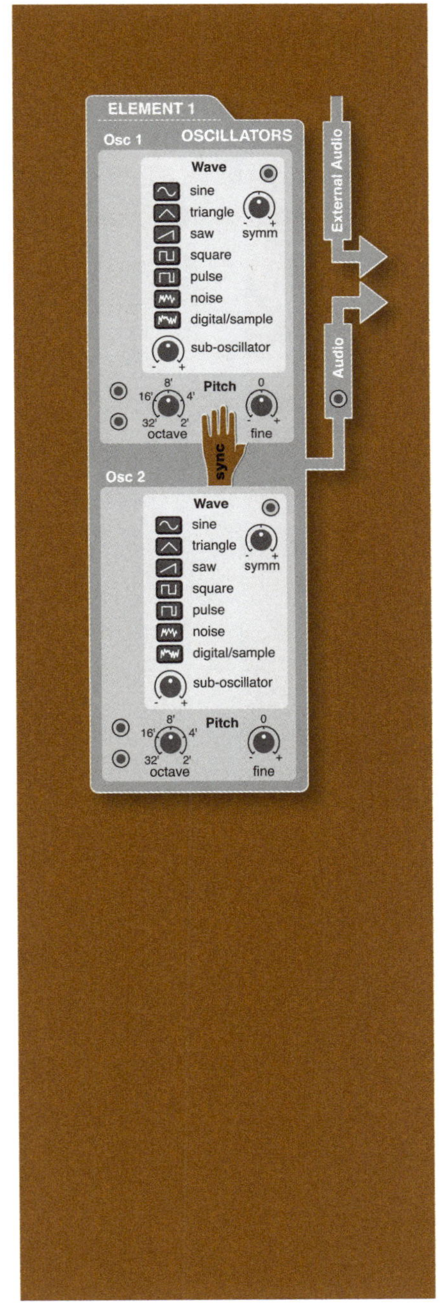

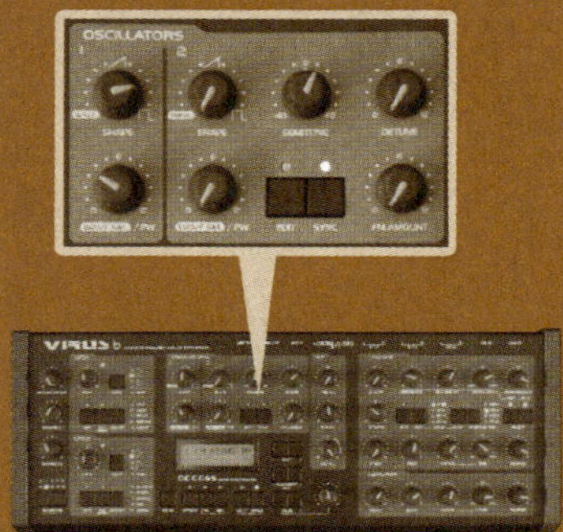

If you have a synth and you press sync at one of the oscillators like the Access Virus on the picture, you might be surprised. Maybe you do not hear any result, or only notice the sound getting a bit tighter. So if you use two oscillators and set one of them into sync, it does sounds more like a one oscillator sound. Not very exciting you might think, what is the use of sync then?

Here comes the fun part. When you press the sync button, oscillator two syncs to oscillator one. The effect is oscillator two stops having independent pitch. It can't detune anymore and instead of that it changes in sound. So... if you change the pitch of the oscillator that is SYNCED.....the sound changes! Hey..., that is exciting again. It gets even more exciting if we take something out of Element-4 and let the pitch change automatically. Element-4 is a chapter for later on, but on the Video 'Oscillator sync', you'll find some extras that will show you what you can do with an oscillator in sync. In the above section I explained sync, but in fact it's called hard-sync because there is also soft-sync. The most common is hard-sync but for instance on the Alesis Andromeda you can select soft or hard sync. Synthesizers using soft-sync try to avoid audible side effects of the hard-sync technique, for instance by reversing the direction of the synchronized oscillator at every new cycle of the master oscillator. But, you will notice that soft is indeed soft and therefore is used less. Regardless, if your synth has the option to switch between hard and soft SYNC, it's always good to play around with it. You never know what kind of nice sounds will pop up. Basically, it's important for you know what sync does and that you can hear the results.

***So remember this:***
- ***If an oscillator is in sync, it has no life of its own.***
- ***Changing the pitch of an oscillator in sync changes the sound, not the pitch.***
- ***To get the best results, change the pitch in a coarse manner or use one of the Element-4 parts to change the pitch.***

***Element-1: 'Oscillator sync'.***

## Oscillators in general

More is better?
As you know, in this training we use only two oscillators in 'The 4 Element Synth' and for purposes of explaining subtractive synthesis, this is sufficient. There are of course synthesizers on the market that have more than two oscillators. Due to the high costs of manufacturing synth technology, in the early synth age, there were synthesizers with only one oscillator (VCO) such as the Roland SH-101.

Roland SH-101
These synthesizers were limited because having two oscillators allows you to build more types of sounds. But if a synthesizer has two or more oscillators do you have to use them all the time? Of course not, it just depends on the kinds of sounds you're trying to create.

Maybe you're thinking the more oscillators I use the phatter my bass sound is going to be? Well, let's take an acoustic example. When you have a bass guitar and you pluck a string it produces a tone. How many strings are active? Yes indeed, only one. Another example is if you play a trumpet: how many mouthpieces does it have? Yes indeed, only one. Take a look at what's happening in the acoustic world and you'll notice that for instance with an electric bass only one string is going to vibrate when you play a note.

These rules of the acoustic world also exist in the world of synthesizers. Which means if you want to make a bass sound basically one oscillator should be enough. If we want to simulate the sound of a trumpet we can do it using only one oscillator. If I use two oscillators I can never get the sound quite right and it doesn't sound like a trumpet. I can turn all the knobs until the year 2060 but it still won't sound like a trumpet.

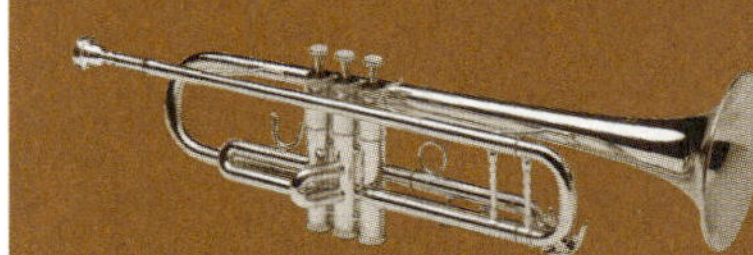

But what about a complete brass section? Yes indeed, we're going to need more than one oscillator to create it. A typical horn section usually consists of a trumpet player, a trombone player and a sax player. So in fact, we're going to need 3 oscillators if possible. And when we're creating string sections, we have to remember that we're going to emulate a group of string players and to do this we need more oscillators in order to create the chorus effect that will thicken the sound. This effect happens naturally with an acoustic string section.

So for creating many different kinds of sounds, it's wise to look at the acoustic world and acoustic instruments. If you have a synthesizer with many oscillators you don't have to use all the oscillators all the time. It depends on the type of sound you're creating and to be sure, a single oscillator sound can be just as impressive as a multiple oscillator sound.

WATCH ONLINE ***Element-1: 'Oscillators in general / More is better?'.***

**Oscillator start phase on today's synthesizers**

A waveform is periodic, it repeats over and over when an oscillator is running. The position within one period is called phase, and is measured in degrees. The starting point of a waveform is called 0 degrees, halfway is 180 degrees, and the end point is 360 degrees which corresponds with the 0 degrees point of the next period.
In the old synth days, there was no control in the oscillator section that let us keep the oscillator waveform from starting at a random point or phase. Even on today's analog synthesizers such as the Andromeda, there's no such control or setting.

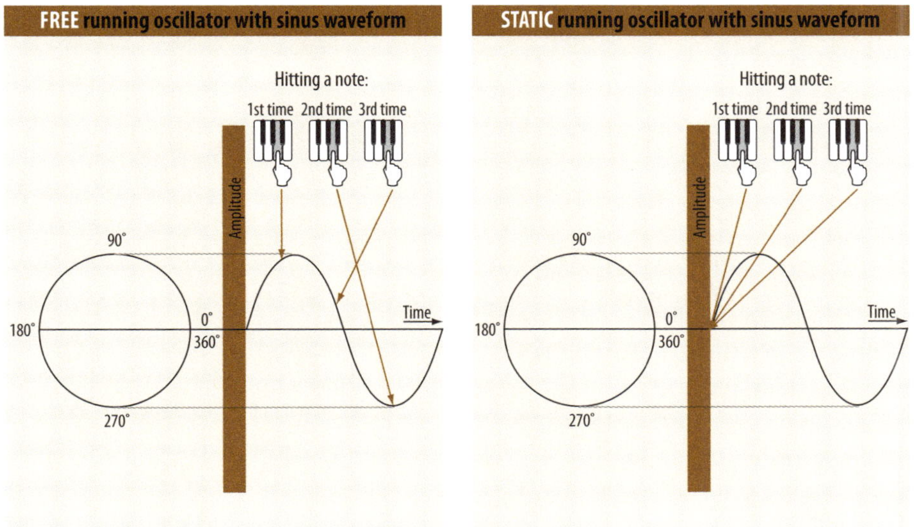


Figure 1.20

However, with digital synths and especially many software synthesizers this control is available. Although a random start phase of the waveform is the best option, it can be handy at times to be able to switch the random start phase 'on' or 'off'. For instance with drum and percussion sounds, setting the free running to 'off' is best.

WATCH ONLINE *Element-1: 'Oscillators in general / Start phase'.*

**What happens if you use two oscillators?**

In the chapter about harmonics we explained to you that the waveform of an oscillator is built up starting from the fundamental and then the harmonics above it. This determines the sound character of a waveform. So, a waveform is in essence, a pile of sinus waveforms. If we use more than one oscillator, the harmonics of both waveforms will mix together. The results of this mixing of harmonics can be described as aliveness inside the sound. When using two oscillators, there are two factors that will determine how the oscillators are going to mix together:

1) the start phase of the oscillator waveform
2) the amount of detuning

When mixing two oscillators, there are two special cases when it comes to start phase. First, when the oscillators are at 0 degrees or in phase, both oscillators are in perfect sync and the waveforms start at the same point. Second, when the oscillators are at 180 degrees or out of phase, the shift between both oscillators is exactly half-way. This means the waveform of one of the oscillators starts halfway the other.

By mixing both oscillators together, the audio that results is a mixture of both of the oscillators.

If one oscillator starts precisely in sync at 0 degrees, the audio doubles in volume as a result of this mixture. (figure 1.21)

If one oscillator starts precisely 180 degrees later, the fundamentals are for instance out of phase and as a result, we end up with no fundamental. (figure 1.22)

All other start phases of the oscillator waveform will give us results that are in between doubling the volume of a given frequency or no sound at all at that frequency. (figure 1.23)

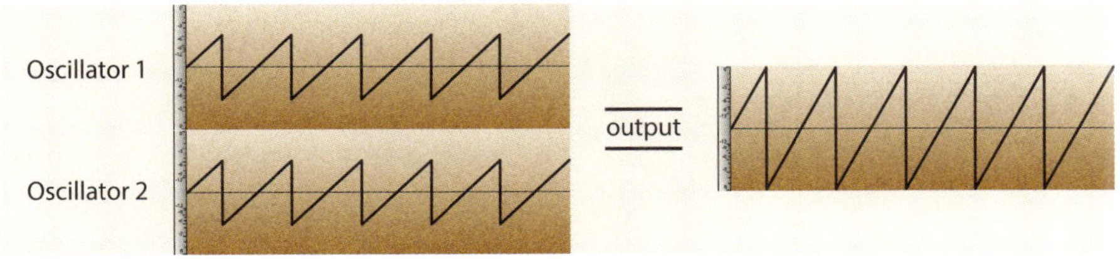


Figure 1.21

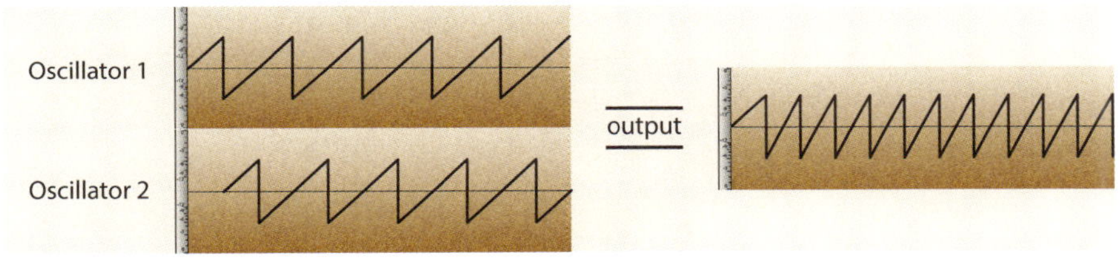


Figure 1.22

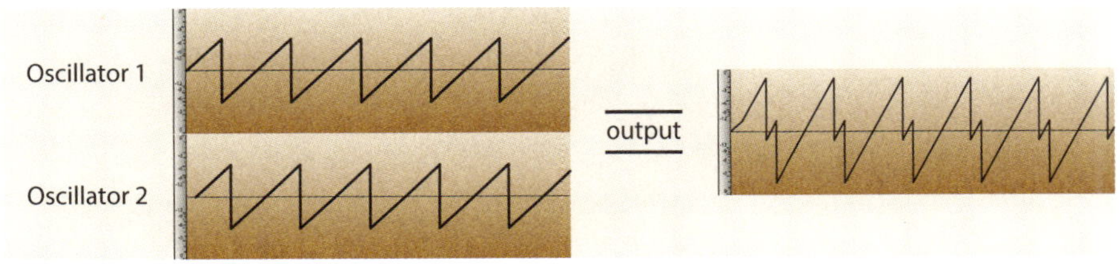


Figure 1.23

Mixing two oscillators.

In the animation on the Video, we've only used one sinus, but if you take a waveform such as a saw waveform, it's made up of many harmonics and all of these frequencies will mix when you use more than one saw oscillator. One of the things that is going on within the sound could be that 1000 Hz is double in volume because the second oscillator is playing the same frequency at 1000 Hz, with a shift of 0 degrees. Another thing going on could be that 500 Hz is gone because of the second oscillator has shifted 180 degrees at this frequency... and this happens only a fraction of time. All of this activity is moving all the time and this is what causes the aliveness in the sound that is the result of using two oscillators.
On top of that, just as the start phase of a waveform has influence on the mixing, detuning has influence as well.

**Problems using two oscillators and bass sounds**

In the beginning of this chapter we explained that it's a good idea to look to the acoustic world for inspiration if you want to build certain sounds. If we take a look at a bass guitar and then we decide to build a bass sound with a synth, using only one oscillator is enough.

Fortunately for us, we play synthesizers, which gives us a lot more freedom. This freedom ends when acoustic rules kick in and you start having problems with your song or mix. This can be especially true if you are trying to use two oscillators for a bass sound.

Okay, but when does the problem start? Because maybe you're already using more than one oscillator for a bass sound. Well, the problem starts if you use two oscillators and have them both set on the same octave setting. Perhaps the sound is not so bad at all, but if you listen carefully you can hear that the bass is not very stable in the lower frequency range. Sometimes it is loud and sometimes soft. What is happening here is that because of this mixing of the frequencies, the fundamental becomes unstable. Something you do not want to have with a bass of course.

The fundamental is also a sinus waveform, the mixing of the frequencies when using more than one oscillator always gives us three possible results: mix, double volume, or anti-phase... which means no sound. And that is exactly what's happening with this bass sound. The fundamental of oscillator one and oscillator two are at the same frequency. When they start to mix, because of the start phase of the waveform and the detuning, the fundamental suddenly sounds twice as loud or... disappears.

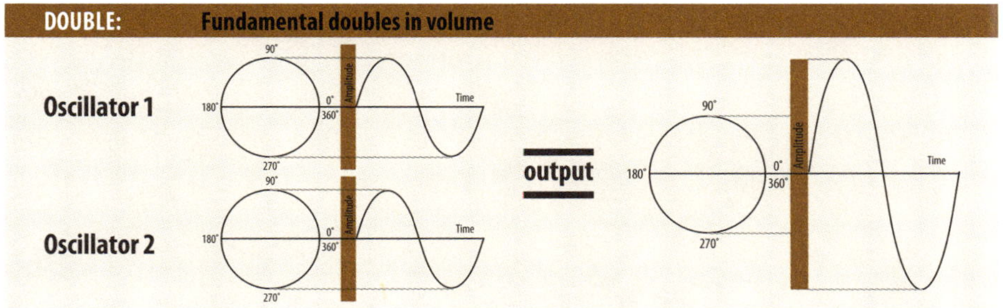


Figure 1.24

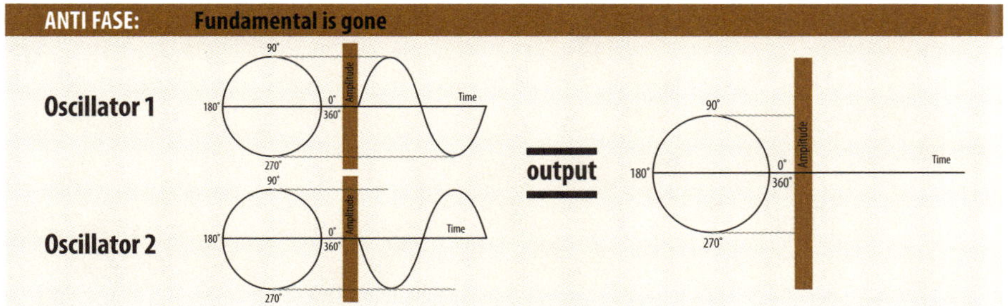


Figure 1.25

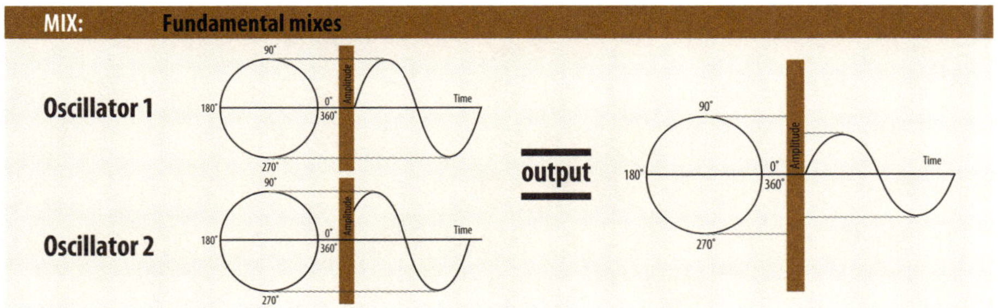


Figure 1.26

To avoid this problem, the best thing to do is to change the octave of one of the oscillators. This ensures that the fundamental doesn't clash with the other fundamental anymore. Another option is to detune both oscillators a lot. This is often done in Drum and Bass music. Because of the heavy detuning the changes in the sound and phases are so fast that it is again a solid and stable sound. But of course, the usage of this type of sound is limited.

*Element-1: 'Oscillators in general / Bass sounds'.*

# Element

# 2

# Element-2 / Filter

**Introduction**

*Introduction*

Check the 4ES chart and look at Element-1. Notice how it connects to Element-2. The audio signal from both oscillators of Element-1 are mixed together and then sent to Element-2.

If you look closely at the chart, you'll see that the Element-2 is named filter. This tells us that the next step in the audio signal route inside a subtractive synthesizer is the filter. Inside the 4ES we have one filter and for explaining subtractive synthesis this is enough. Of course, in the real world, two or more filters are possible as well as variations. But keep in mind these are only variations and for explaining the basic concepts we do not need them.

We have learned that the basis of each sound begins with Element-1, which are called oscillators. With these we generate the raw audio material that we will further sculpt with the other synth tools, which in this case are called filter or Element-2. Another green connection you see is called external audio. This means that external signals from your mixer or another instrument can also be sent into Element-2/ Blue.

Let's start at the beginning: Element-2 filters the harmonics of the audio signal that comes from the oscillators. That's why this type of synthesis, which is the basis of our training here, is called subtractive: we are taking out parts of the whole sound being generated by the oscillators using the filter. Of course, subtractive means not only that the filter is used in a static way, but can also be used in a very dynamic way using tools from Element-4.

Okay, here's what's important to know about this filter. A synthesizer filter of a subtractive synthesizer has two dynamic parameters which we call cutoff and resonance. These are parts you can change dynamically by using your hand (turning the knobs) or by using parts from Element-4 to do it automatically. Therefore, in the 4ES chart the cutoff knob and the resonance knob are the only ones that have modulator inputs.

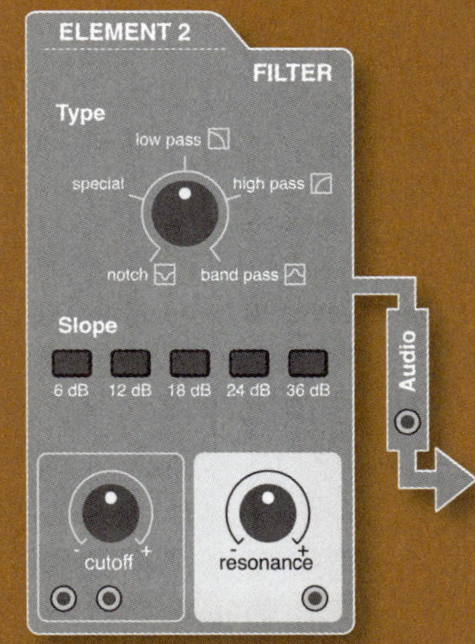

The settings above the cutoff knob and the resonance knob inside Element-2, such as type and slope are not dynamic and only determine how the filter works. For example, with these controls we can set a filter to work like a low pass filter with 24B slope, or we can change the settings and create a band pass filter with 12dB slope. But these controls can't be changed dynamically with Element-4 parts and you can conclude this because you do not see any modulator inputs on the chart. So in reality this makes synth life easier for us: only two parameters are dynamic and the other ones are static!

***Remember this:***

- ***Element-2 alters the sound that comes from Element-1/oscillators and is called the filter.***
- ***Element-2 or filter has two dynamic parameters, cutoff and resonance, that can be controlled by hand or by Element-4 parts.***
- ***Element-2 or filter settings type and slope are static and are used to set up how the filter works.***

***Element-2: 'Introduction Element-2 part 1 and part 2'.***

## Filtertypes

*Filtertypes*

In the introduction to the filter section on the Video, we showed you that many knobs on your synth panels are not really filter parts. But if you look at the 4ES chart, specifically the Element-2 again, you will see that type and slope are really filter parts.

So if you are unsure about certain parameters that you see on your own synth, just grab the 4ES and you'll find the answer. Before we continue we have to look at some basic principles of filters.

Filters are categorized into types by the way they filter a sound. A low pass filter or LPF does exactly what its name implies: it filters high frequencies, and passes low frequencies. The exact frequency the filter changes from passing audio to filtering audio is called the cutoff frequency.

A high pass filter or HPF is the reverse of a low pass filter: it filters low frequencies, and passes high frequencies. Again, the frequency the filter changes from passing to filtering audio is called the cutoff frequency. These two filters, LPF and HPF, are the most common filters in synthesizers.

Apart from these types we also have band pass filters, which only pass a certain frequency band and filter the rest of the audio, and notch filters which are the reverse of band pass filters: they filter a certain frequency band and pass the rest of the audio. In these last two cases the center frequency is called the cutoff frequency.

*Element-2: 'Filter types'.*

*Cutoff frequency*

**Cutoff frequency**

In the introduction of Element-2 we told you that cutoff and resonance are the only dynamic parameters available. We can conclude this fact if we look at the 4ES chart and Element-2.

The cutoff and resonance have inputs and that means they can be controlled by parts from Element-4, which could be an envelope or the keyboard and so on. But what is this cutoff parameter, and what does it do exactly?

Figure 2.01

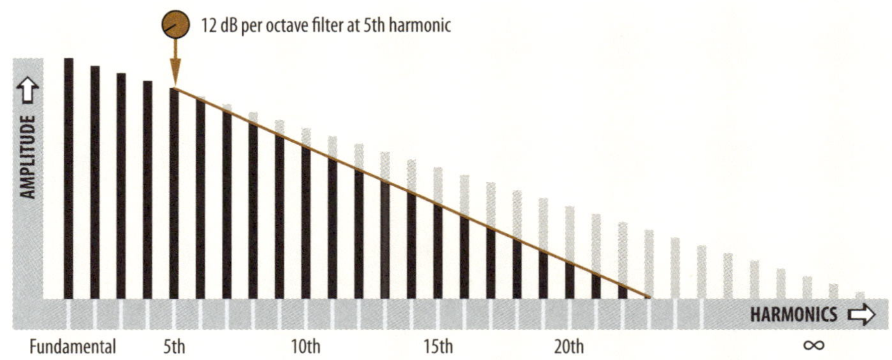


Saw waveform showing the harmonics with cutoff frequency set at the 5th harmonic using a 12dB low-pass filter.

The cutoff knob, or as it's sometimes called, the cutoff frequency, sets the frequency point from where the filter will start to cutoff. So in this case the name of the parameter describes the function. Exactly what will be cutoff depends on the type of filter you're working with.

***Remember this:***

- ***The cutoff knob, or what we sometimes call the cutoff frequency, sets the frequency point from where the filter starts to cut off.***
- ***The cutoff parameter is always active in every regular sound.***

***Element-2: 'Cutoff frequency'.***

## Filter slopes

Apart from a type and a cutoff frequency filters have a slope. This defines how hard the filter works starting at its cutoff frequency.

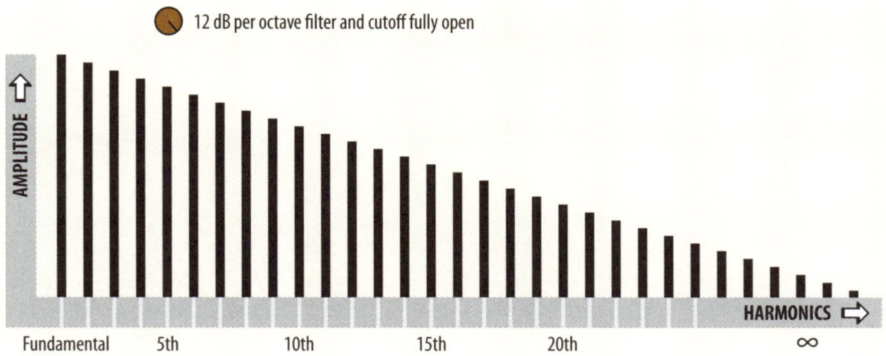


Figure 2.02

Saw waveform showing the harmonics with cutoff frequency fully open using a low pass filter.

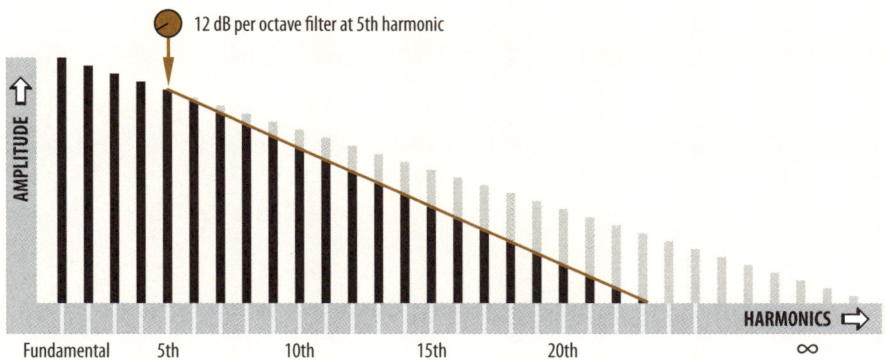


Figure 2.03

Saw waveform showing the harmonics with cutoff frequency set at the 5th harmonic using a low pass filter.

Harmonic 10 is now 12dB softer in volume.

The higher the harmonics in a signal get from the filter's cutoff frequency, the more the filter attenuates the harmonics, and this is usually notated in dB per octave. An octave is a harmonic or frequency multiplied by two. So the 10th harmonic is the octave of the 5th harmonic.

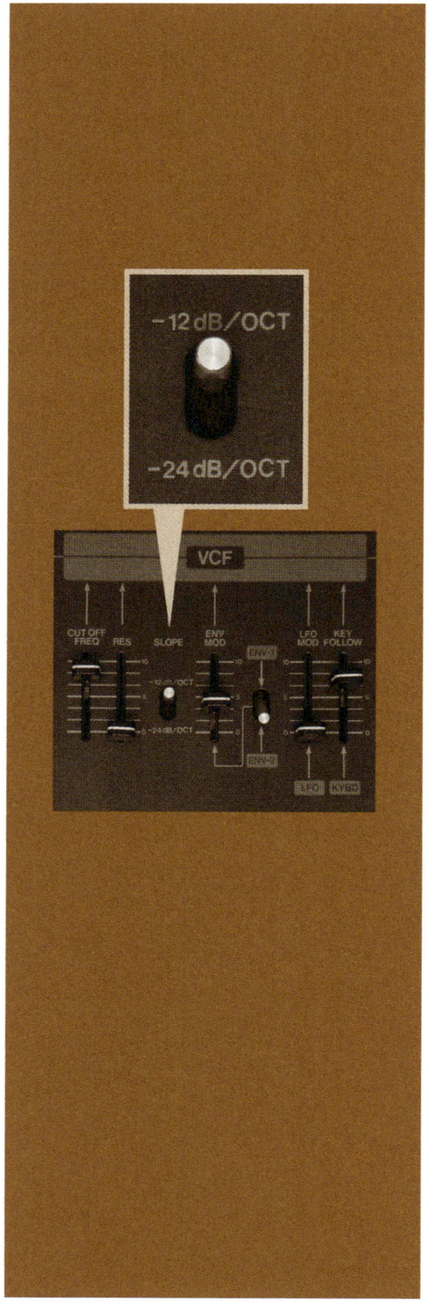


Figure 2.04

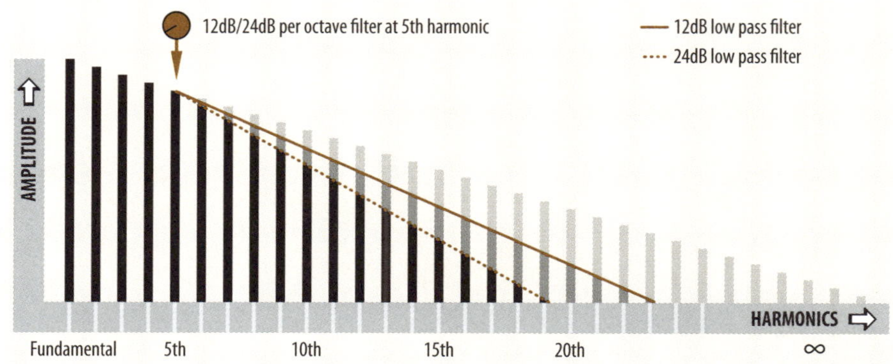


Saw waveform showing the harmonics with cutoff frequency set at the 5th harmonic using a 12dB and 24dB low pass filter. Harmonic 10 is now 12dB softer in volume or 24dB softer in volume, depending on the slope.

A filter which drops its signal 24dB down for every octave away from its cutoff frequency is more steep than a filter which drops its signal only 12dB per octave. So a 24dB per octave filter more effectively filters the harmonics in an audio signal than a 12dB per octave filter, and therefore has a different sound.

You may notice filter slopes always being a multitude of 6dB. Due to the technical design of filters, they use slopes that are a multitude of 6dB. These are called poles: a filter with a 6dB per octave slope is called a one pole filter, a 12dB per octave filter two pole, 18dB per octave three pole, etcetera.

## Resonance

*Resonance*

In the introduction of Element-2, we showed you that according to the 4ES chart, the cutoff and resonance are the only dynamic parameters. Inside the filter of a synthesizer resonance is a tool to increase the frequency at the cutoff frequency point. So if I set the cutoff to a certain position, that frequency point is attenuated or emphasized by the resonance.

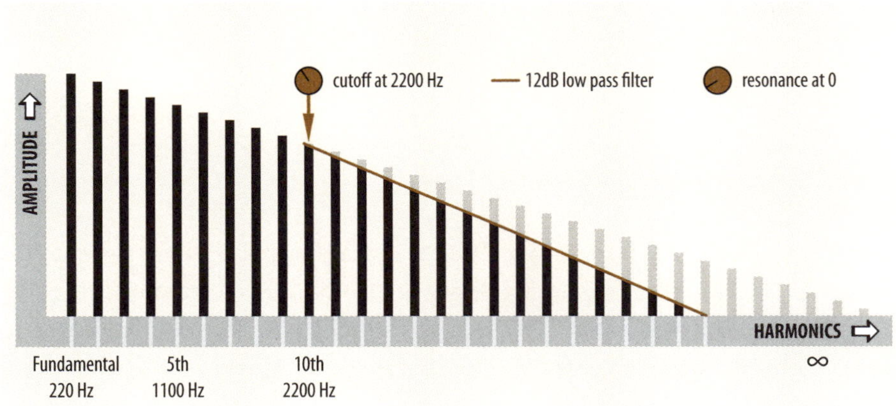


Figure 2.05

Saw waveform with cutoff set at 2200Hz and resonance not active.

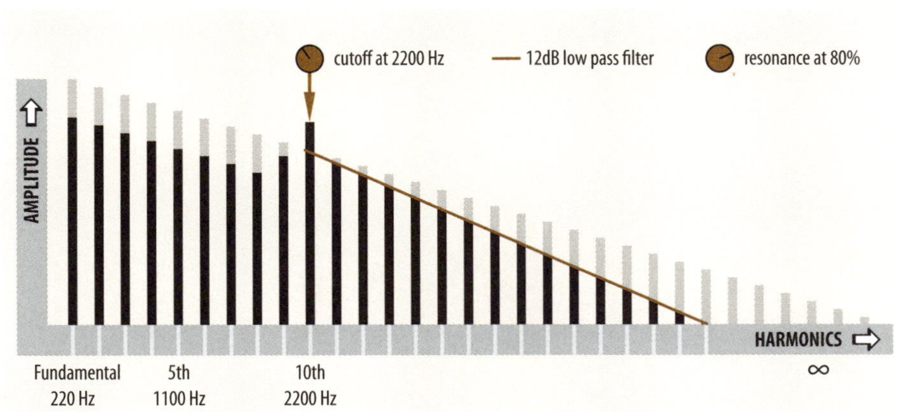


Figure 2.06

Saw waveform with cutoff set at 2200Hz and resonance at 80%.

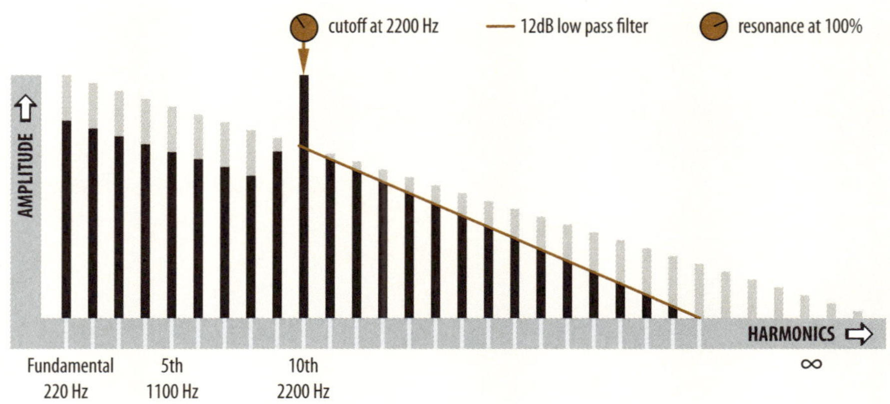


Figure 2.07

Saw waveform with cutoff set at 2200Hz and resonance at 100%.

As you can see, there is a peak that starts to rise at the point where the cutoff is positioned. So if the cutoff frequency is set at 2200 Hz, then that frequency is going to be amplified by the resonance. This is accomplished by using a feedback loop in the filter, just like what you have sometimes heard if you've ever sang in a band and the microphone was feeding back.

Once you start opening the resonance on the synth, the feedback loop starts to happen inside the filter. The result is that the frequency at which the cutoff is set starts to increase. Once the feedback is at maximum a tone is produced. This tone is the pure frequency of the cutoff frequency and we call that a sinus waveform. When you hear the term 'self-oscillating filter' this is what is happening.

***Element-2: 'Resonance'.***

## Resonance and its uses

As we explained that resonance attenuates the frequency that is set by the cutoff frequency. The overall sound of the filter changes, because if one frequency gets louder the other frequencies get softer. So the sound of the filter changes in a way that is very special and can't be found in any other acoustic instrument.

If I open the resonance just a bit, you can hear that the sound changing. If I change the cutoff frequency, the frequency that is being attenuated by the resonance moves as well. So resonance inside a synthesizer can move and change, very different compared to an acoustic instrument! It gets even more exciting if we apply something from Element-4 on the cutoff frequency, like an envelope. Then the cutoff frequency starts to move automatically. When such a device like a filter envelope is utilized, the resonance can add extra juice to the sound.

*Element-2: 'Resonance uses'.*

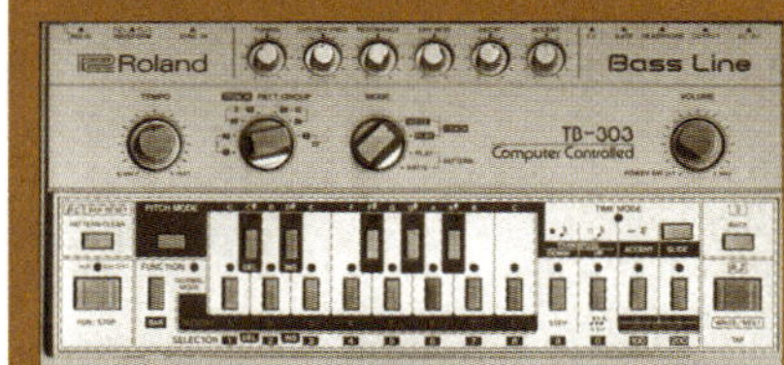

*Filter types*

## Filter types

### Low-pass filter

As we said in the introduction, filter modes are not available on every synth. When I pick a MiniMoog or Sequential Prophet-600, Roland Juno-60, Waldorf Pulse or Microwave, they all have only one type of filter available. This should be no real problem because as you know these are all great synthesizers. A synthesizer is much more than an endless amount of features. The sounds a synth is capable of producing and also what you can do with it, that's what really counts.

So most synthesizers have only one type of filter and you can bet it is a low-pass filter. Most of the time this low-pass filter is a 24dB filter, but there are also synthesizers with 12dB filters. A low-pass filter allows low frequencies to pass through the filter, while it filters the high frequencies and harmonics that come from Element-1. So if you have opened the filter by opening the cutoff knob, all frequencies are passed through. Once you start closing it, the high harmonics will begin to be cut off at a rate of 24dB.

If we look at Element-2 again, we can see that we have several options in slope: 6db, 12dB, 18dB, 24dB and 36dB. A low-pass filter with 6dB is rare and is only used in software synthesizers and samplers. The most common are the 12dB and 24dB low-pass filters. Sometimes an 18dB low-pass shows up, which they say was used on the Roland TB-303. The low-pass filter is a basic need for any subtractive synthesizer, so of course, it's the most important filter type.

*Element-2: 'Low-pass filter'.*

### 12dB low-pass filter

In this section, we show you images that display how a 12dB low-pass filter cuts off the high harmonics. To show you the difference between a 24dB and a 12dB filter we've also included a images that shows both filter curves. In the first image, the cutoff frequency is at 2200Hz.

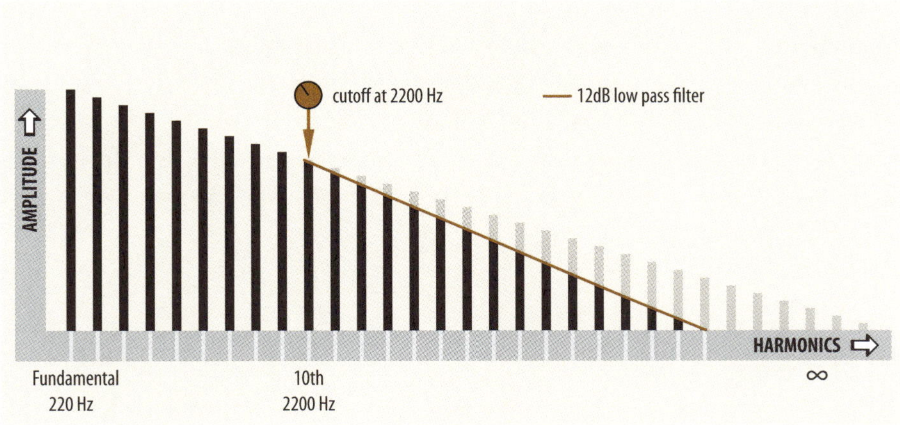


Figure 2.08

12dB low-pass filter, cutoff frequency is at 2200Hz.

What's happening is that the high harmonics are being cut off above the cutoff frequency point with a slope of 12dB per octave. They decay at a rate of 12dB per octave. So, for example, harmonic 20 becomes 12dB lower in volume if the cutoff frequency is set to harmonic 10. The next image of the 12dB and 24dB Low Pass filter shows the difference between them. Compared to the 24dB filter, the upper harmonics will remain louder with a 12dB filter.

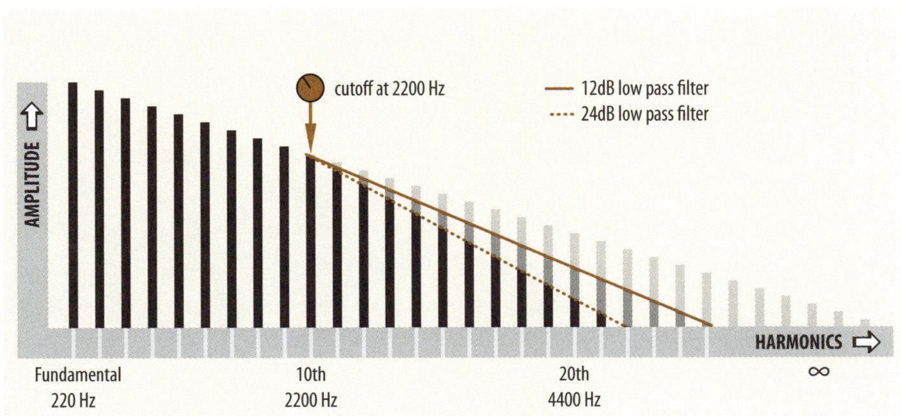


Figure 2.09

The difference between a 12dB and a 24dB low-pass filter.

But some of you might think now: "If I close the 12dB cutoff frequency more or maybe even twice as much, then it will be the same as the 24dB filter"? And yes indeed, you can turn the 12dB cutoff frequency more and then the filter will close much more, but no... the sound will never be the same as a 24dB Filter! The way the Filter reacts depends on the position of the cutoff frequency and the slope. Look at the 3rd image.

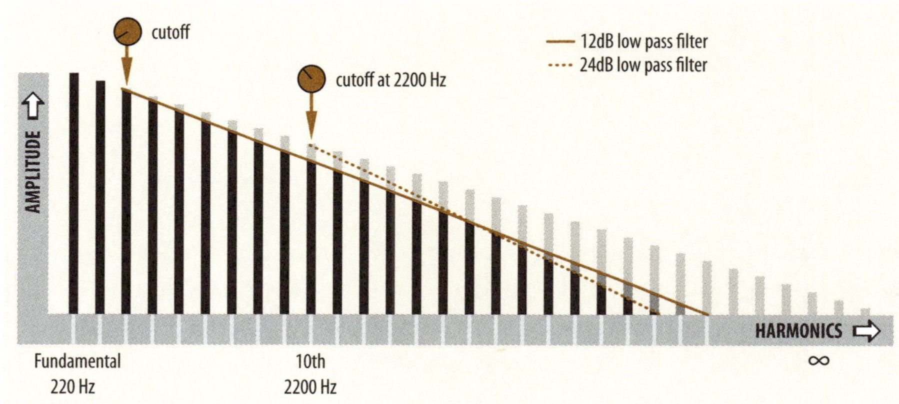


Figure 2.10

12dB filter cutoff closed twice as much, compared to the 24dB filter.

You can see that the 12dB filter cutoff is closed twice as much compared to the 24dB filter. This is an attempt to make it sound like a 24dB filter. You can see very clearly the harmonics after filtering are very different. So even if you close the 12dB filter twice as much as a 24dB filter, the sound will be different. This has to do with the fact that we have 12dB per octave versus 24dB per octave attenuation.

So you can see now that a synthesizer is a lot different from an EQ on your mixer! The 12dB filter kicks in less sharply and will allow the upper harmonics to remain louder compared to a 24dB LP filter. The results can be described by saying that the 12dB filter is more open sounding.

Therefore, the 12dB filter is great for silky pad sounds and open sounding synth sounds, and is often used in rompler players, workstations and sample libraries.

### High-pass filter

The high-pass filter is less common and some older synthesizers don't even have one. The high harmonics and frequencies are passing through this filter unchanged. The filter type also determines the function of the fader.

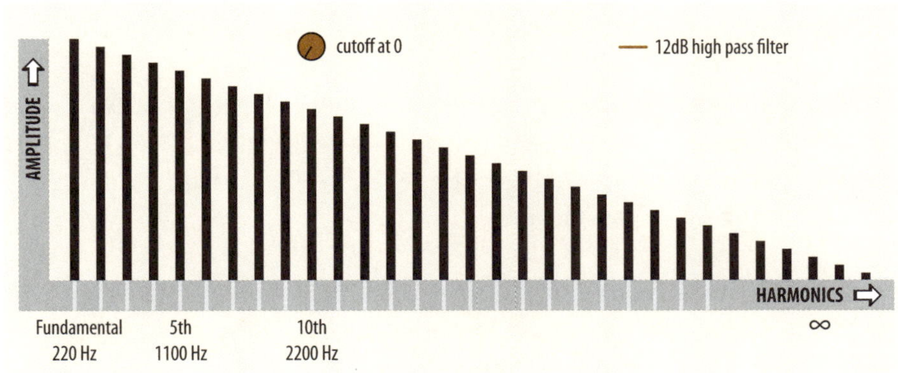


Fader or knob closed and high-pass filter open.

Figure 2.11

With the fader or knob closed, the high-pass filter is open or in other words, not filtering, because the high-pass filter is filtering the low frequencies and harmonics. Therefore it's called a high-pass filter. If you open the high-pass filter, it starts filtering by cutting off the low harmonics and frequencies. The higher you move the fader, the less of the lower harmonics and frequencies you'll be able to hear.

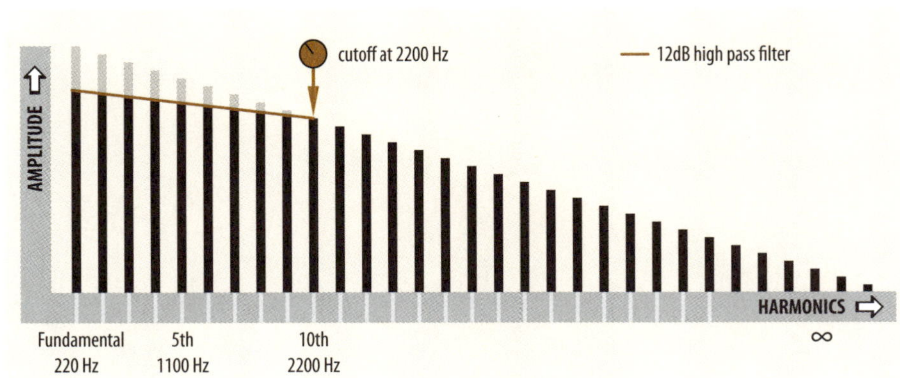


Figure 2.12

12dB high-pass filter cutting off the low harmonics.

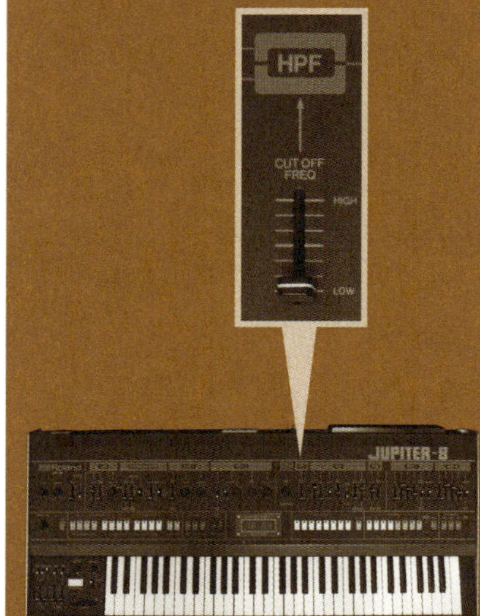


What can we learn from this?
We learn that just opening and closing the cutoff frequency knob has nothing to do with opening and closing the filter. It depends on the filter type to determine what is open and what is closed. If we look at Element-2 again, we can see that we have several options in slope: 6db, 12dB, 18dB, 24dB and 36dB. A high-pass filter with 6dB is very handy but only used on software synthesizers and samplers.

The most commonly used are the 12dB and 24dB high-pass filters. Just like the low pass filter, the higher the dB value, the harder a filter cuts off. So a 24dB high-pass filter is much stronger than a 12dB filter. The high-pass filter is used quite differently in practice than the low-pass filter.

A synthesizer can function without a high-pass filter, but not without a low-pass filter. Think of it more like a corrective filter used to avoid audio problems. Of course, there's also other cool things possible using a high-pass filter.

WATCH ONLINE ***Element-2: 'High-pass filter'.***

**High-pass filter as a corrective filter**

We talked in the Element-1 chapter about harmonics and how they mix if you're using more than one oscillator. When we're making music, we're always combining several sounds together and sometimes conflicts crop up.

For instance, in a mix we use a bass sound and a sequence sound playing at the same time. Of course, not only within the bass sound we have a fundamental and harmonics, but also in the sequence sound. The low frequencies within the sequence sound can conflict with the bass sound. You can avoid this conflict by opening the high pass filter a bit on the sequence sound. You can hear that the bass adds more air into the mix and the whole mix now sounds more open. If you solo the corrected sound, you may conclude it sounds thinner. And indeed, if you pull the high-pass filter down again, the sound will become fuller again. But in combination with the bass you won't perceive that the sound is thinner because of the high-pass filter. In fact, the overall sound improves.

Of course, you could also use EQ on your mixer to roll off the bass amount of the sequence sound. But if your synthesizer has the option to use a high-pass filter, why not use it?

WATCH ONLINE *Element-2: 'Corrective high-pass filter'.*

## Band-pass filter

The first two filter types we've discussed so far have been the low-pass filter and the high-pass filter. If you combine these two filters in a series and use a similar setting for the cutoff frequency for both filters you'll in fact have the same thing as a band-pass filter. The band-pass filter cuts off the high harmonics and the low harmonics and leaves a certain band of harmonics that passes on through the filter. Hence its name, band-pass, which means that only a certain band of harmonics are allowed to pass through the filter.

Figure 2.13

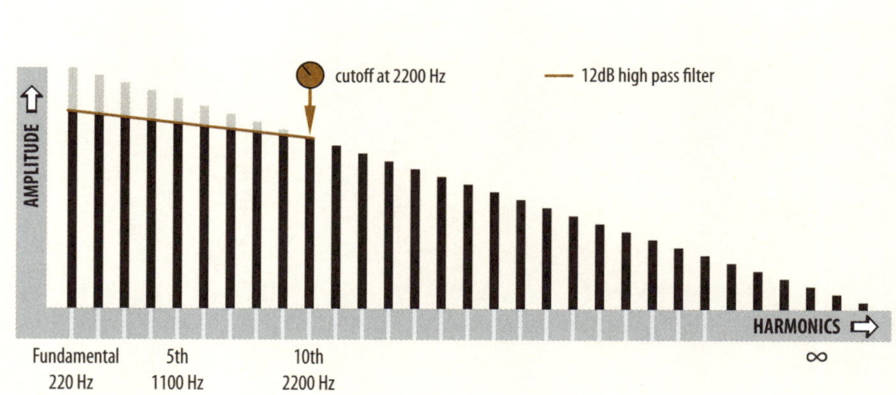


High-pass filter set at 2200 Hz.

Figure 2.14

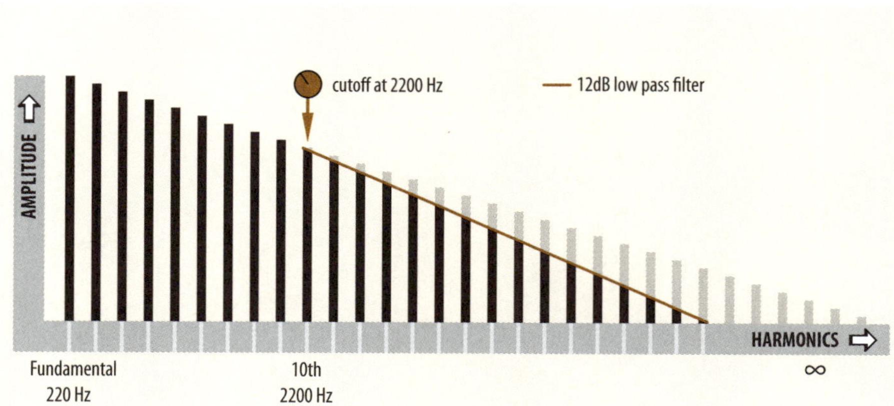


Low-pass filter set at 2200 Hz.

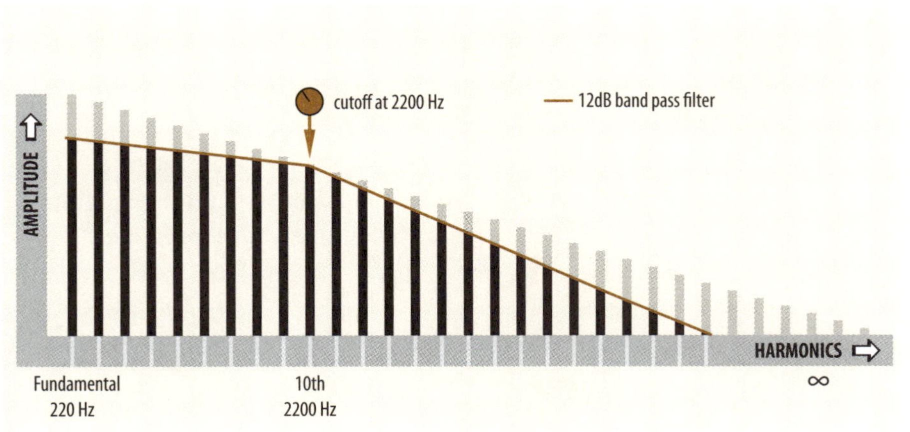


Figure 2.15

Low-pass filter and high-pass filter, layered together.

Figure 2.15 shows a low-pass filter and high-pass filter that have been layered together. The second image is a band-pass filter. So the band-pass filter is built up using two filters in series, a high-pass and a low-pass filter.

The famous Korg MS20 synth has both a high-pass and a low-pass filter. If you set both of the cutoffs at the same position, it is set up as a band-pass filter. But if you want to use this band-pass setup on the MS-20, you'll need to keep the cutoff frequency of both the filters at the same frequencies all the time. So if you change the low-pass filter cutoff, then you have to also change the high-pass filter cutoff as well and set it to the same position or frequency.

Inside many synthesizers a band-pass filter appears as one filter type. The advantage of having it as one filter type, is that you only need one cutoff to internally control both the low-pass and the high-pass filter. But as you've seen with the Korg MS-20, you can make your own band-pass filter if you have a high-pass and low-pass filter in a series inside your synth, in case a dedicated band-pass filter is not available.

We can say that the band-pass filter has no neutral position, it's always filtering. That is logical because two filters are used to make up a band-pass filter and both have the same cutoff frequency position. The band-pass filter is, as said before, a high-pass filter and a low-pass filter in series. Both these filters have a neutral position.

The neutral position of the high-pass filter is achieved by turning the cutoff knob fully to the left. The neutral position of the low-pass filter is achieved by turning the cutoff knob fully to the right.

Well, we've said that a band-pass filter has these two filters in series, and they are both set with the same cutoff frequency. So, to simulate a band-pass filter, I'll need to move the high-pass to the right, which now makes the filter active instead of neutral. And likewise, I'll need to move the low-pass to the left, which makes it active instead of neutral. So if you bring the cutoff together from both filters in order to make it a band-pass filter it means that it's always going to filter.

This is all good, but what is the musical significance of the band-pass filter? It is a filter type that dramatically changes the sound, because only a specific band of harmonics coming from Element-1 are passed through. This makes the band-pass filter quite useful for creating sweep sounds and other typical synth sounds.

WATCH ONLINE ***Element-2: 'Band-pass filter'.***

**Notch filter**

The first two filter types we've discussed were the low-pass filter and the high-pass filter. If you combine these two in parallel and set a gap between the cutoff frequency positions on both filters, you have now created the same thing as a notch filter. On the next page (figure 2.16) you'll see an image that shows this gap.

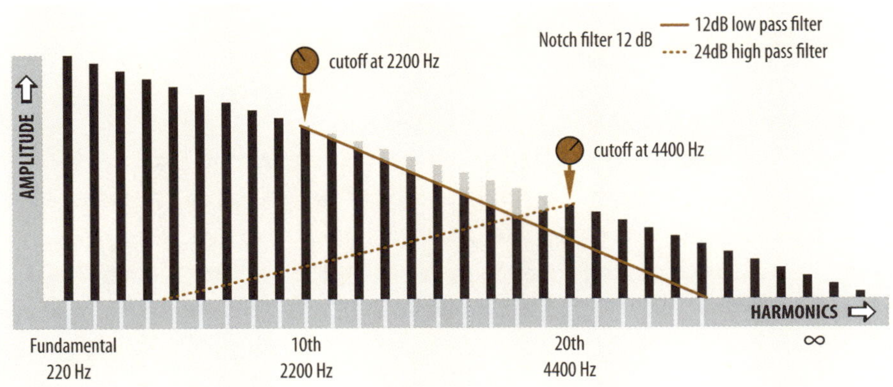


Figure 2.16

Notch filter.

The notch filter is also commonly known as the band reject filter. In terms of sound, it's a reverse band-pass filter, it filters a specific range of harmonics. The size of this range of harmonics is determined by the gap between the low-pass filter cutoff and the high-pass filter cutoff. Most synthesizers use a default setting for the gap. Note that the high-pass filter will always be higher in cutoff frequency than the low-pass filter, otherwise there will be no gap.

*Element-2: 'Notch filter'.*

**Special filter**

On many of today's digital synthesizers, there are special filters which don't fit in these standard categories. The most popular are the comb and vowel filters, which we can find in E-mu synthesizers and samplers, and in software synthesizers like BLUE and Predator.

*Element-2: 'Special filter'.*

*Cutoff Frequency modulation*

## Cutoff Frequency modulation

Introduction

In the previous section we explained to you what synthesizer filters are. We showed you that many of the parameters and knobs that you see on the front panels of synths are not really parts of the filter. Manufacturers make synthesizers easier to program and edit by adding other components inside the filter section. This can be confusing when you're trying to understand the basic principles of subtractive synthesis. So, if we look at the 4ES chart, bear in mind the elements that are separate may have a different appearance and will be mixed together on your synthesizer panel.

With the invention of the MiniMoog, which was one of the very first stage and performance synthesizers, they started to combine the parts that made synthesizers more suitable for performance and comfortable to play and program. Instead of using real wires to connect the elements, like the ones that were used on modular synths, the connections where pre-made at the factory. So if we translate the MiniMoog to the 4ES chart we can see that the filter's panel version has Element-2 parts which are cutoff and resonance, and Element-4 parts which in this case are an envelope with attack, decay and sustain settings.

So, the panel shows filter as its name....and from the 6 knobs and 3 buttons we see on the MiniMoog only 2 are really filter parts! All the other parts we just mentioned on the MiniMoog are Element 4 parts and these parts change or modulate the cutoff frequency. This happens internally because you don't see the cutoff knob actually moving.

Now, let's quickly repeat this by using the MiniMoog as an example:
- cutoff and resonance are the only real filter parts on the filter panel
- envelope amount, attack, decay and the other parameters are cutoff frequency modulation parts.

Compared to the 4ES chart:
- cutoff and resonance are Element-2 components.
- other knobs and functions are Element-4 components and carry out the function of cutoff modulation.

Good.... this is a lot of information to cover just to introduce you to the cutoff frequency modulation. So, let's recap for you a few important things that we covered in the other sections about Element-2 and add it to our list of new facts.

***Remember this:***

- ***cutoff, resonance and filter type are the only real filter parts.***
- ***cutoff frequency is always used and is the central parameter of the filter.***
- ***all the other parts that you see on your synthesizer filter panel are Element-4 parts.***
- ***all these Element-4 parts modulate the cutoff frequency.***

Element-4 is very important when used in combination with a filter. When we are faced with only having cutoff, resonance and filter type to work with in a static way we are not going to be able to get our sounds up to the level we really desire. We need the Element-4 parts such as envelopes and LFO's to make the synth alive and kicking!

Remember at the very start of this training we told you that Element-4 is a vital key in making a synthesizer work. I will repeat the function of Element-4 again for you: it makes the other elements work and add dynamic changes. So if we look at the 4ES chart again we can see that cutoff frequency and resonance both have inputs. From Element-4 we can connect other things to these parameters of the filter.

***Element-2: 'Cutoff frequency modulation'.***

**Cutoff frequency envelope modulation**

The major Element-4 part for the filter is called envelope. What is an envelope? And what does it do? We are inside the section of Element-2 and in the middle of explaining the filter, but as I mentioned at the very start of this booklet we can't talk about Element-1, Element-2 or Element-3 without talking about Element-4.

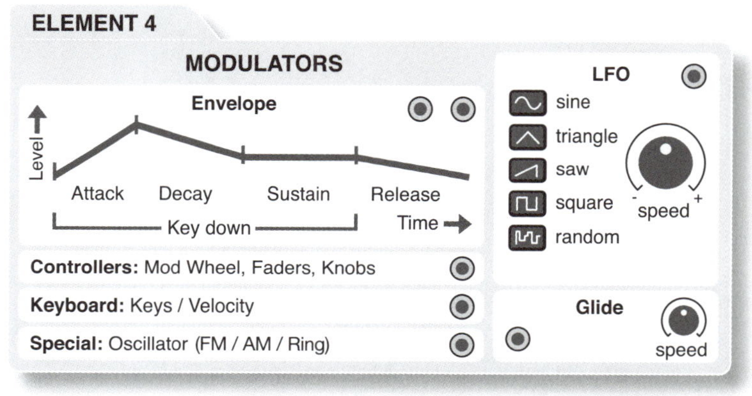


And the different synthesizer panels show us again and again that the envelope plays a crucial role inside the filter section. Right now let's do this as briefly as possible, but you can dig deeper into the subject further on in Element-4 section.

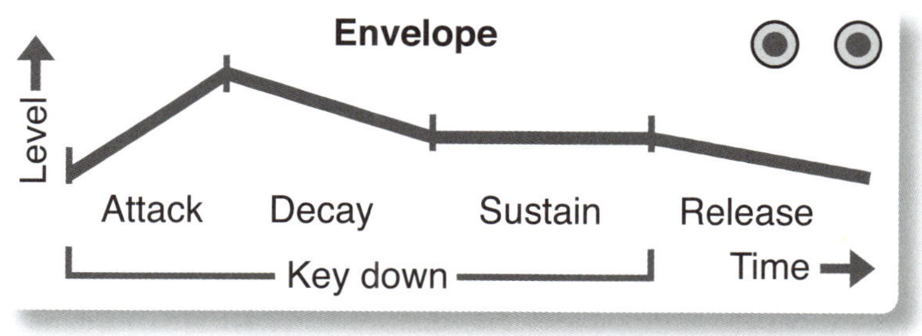


The basic envelope parameters are attack, decay, sustain and release. If you look at the 4ES chart you can see them in the Element-4 section as well.

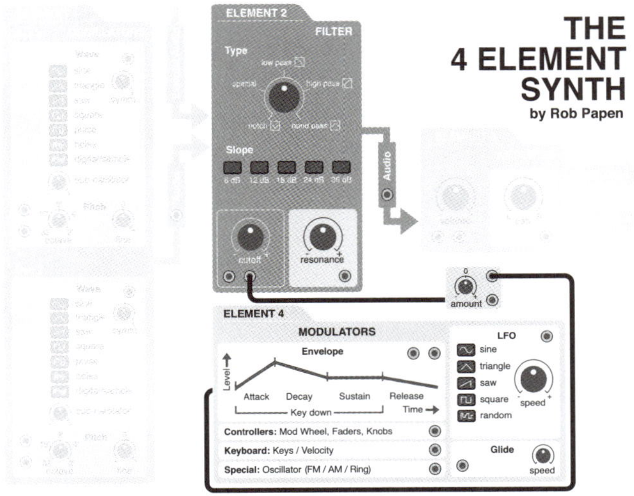


The next step is to connect the envelope to the cutoff frequency. But the connection of the envelope to the cutoff frequency is never a direct one, as there is always an amount knob in between the connection. The amount knob is needed to control how much of the envelope we want to apply to our sound. This is because you can apply a lot of envelope or a little, it depends on the sound you are trying to create. Now back to the envelope itself.

***An envelope with attack, decay, sustain and release is a time-based module inside a synthesizer. It is connected to the keyboard and once you hit the keyboard it starts to work.***

You can apply this movement to the cutoff frequency by using the envelope amount knob. By changing the settings of the envelope you change how it affects the cutoff frequency over time. This means that you can create sounds simply by editing the envelope parameters.

***Element-2: 'Using envelope modulation'.***

**Cutoff frequency envelope modulation and their secrets**

There are a few things that you have to keep in mind when using the envelope to modulate the cutoff frequency. In fact, there are three points that have great value if you're going to modulate the cutoff frequency with an envelope. These three points will give us endless options in sound, so we need to know them to get the max out of your synthesizer.

*[1] Envelope and its settings*

On the panel of a synth we usually have the parameters attack, decay, sustain, and release for the envelope.

*[2] Cutoff frequency and it's setting*

On the panel of a synth we have the cutoff frequency as the central knob of the filter, but it is also the initial setting. In other words, it will set the initial point from where the modulation, and in this case, the envelope starts.

*[3] Envelope amount*

On the panel of a typical synth we see the envelope amount knob, and this controls how much the envelope is applied to the cutoff frequency. If this knob is closed the envelope will have no effect on the sound, but if we open it fully, the envelope will work at the maximum on the cutoff frequency.

With these three interconnected points we now have a ton of options!

Keep these three points in mind if you're going to use the envelope. First, the cutoff frequency setting is the initial setting. Second, the envelope amount setting sets how deeply the envelope modulates the cutoff frequency And third, the settings of the envelope itself. Another important point about envelope modulation of the filter. Many producers and synth players use only the cutoff knob when editing sounds. But remember, there is also the envelope amount and the envelope settings that can change the sound. And not just a little bit!

### Cutoff frequency keyboard (note) modulation

The keyboard as a modulator is not only used to control the pitch of the oscillators. It can also be used to control the cutoff frequency of the filter. It must be an important component, because we can find this keyboard modulation on almost all filter panels on many synthesizers.

The name for cutoff modulation using the keyboard is often called different things. On the Roland Jupiter-8 it is called key-follow while on the Alesis Andromeda it is called key-track. But both function the same.

The keyboard modulates the cutoff frequency and this means that the cutoff frequency changes internally according to the note you play. Why did they add this? To explain this, let's look at the acoustic world.
If we play an acoustic instrument, the amount of harmonics is different for high notes as compared to low notes. Our human ear expects this to be the same with a synthesizer. So if we use a synthesizer, then the higher harmonics have to be louder on the higher notes than they are on the lower notes.

But there is also a technical reason. By design, the cutoff frequency of a synthesizer is static with a fixed value, while the notes we play usually change all the time.

For example, if you play a sound without keyboard cutoff modulation it sounds good in the lower keyboard range. But if you play it in the higher keyboard range, it sounds too dark. It's so dark you'd even think that the filter is closed! But that is not the case. The cutoff remains in the same position. But our ears expect more harmonics in the higher keyboard ranges, so you have to fix this. You fix it by using the key-follow or keyboard applied to the cutoff frequency. If you play this sound again, you can hear now that by adding a bit of key follow it has a more natural sound, even in the higher keyboard ranges.

Let me show you also using these 3 pictures.

Figure 2.17

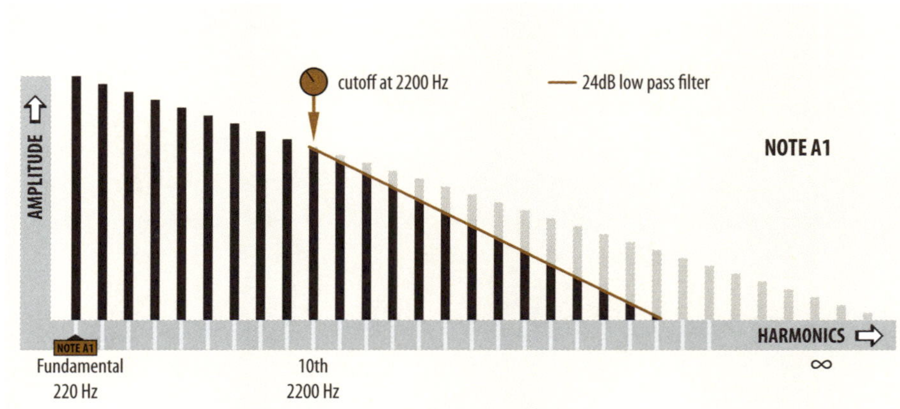


A1-note playing. Correct setting of the cutoff frequency for a particular sound.

At this A1 note, the setting of the cutoff frequency for a particular sound is correct.

Figure 2.18

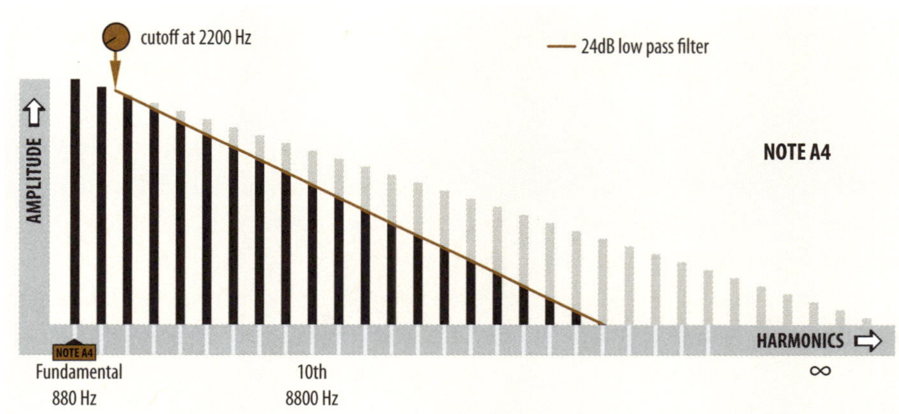


A4-note playing. Wrong setting of the cutoff frequency for a particular sound.

The filter and its cutoff frequency are static. If you play higher notes, the whole filter setting is wrong because the note's fundamental and its harmonics changed. Look at the image with note A4 played. You can see that the harmonics are much more attenuated than in the first image. This is because the cutoff did not move along with the changed pitch. Result is that the sound is incorrect and the sound is more darker than it should be.

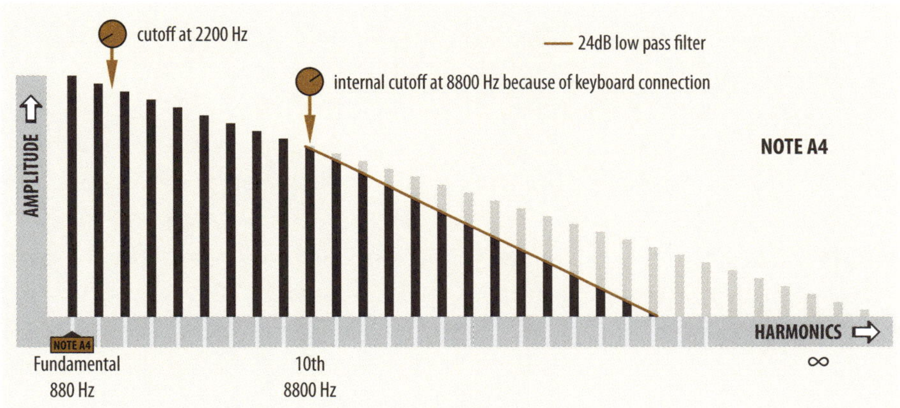


Figure 2.19

Cutoff frequency is now controlled by both the filter setting and the keyboard.

Cutoff frequency is now controlled by both the filter setting and the keyboard. The sound is now correct at this given note A4. The filter knob at the panel of the synthesizer remains at 2200 Hz, but internal the setting is changed by the connected keyboard to 8800 Hz and starts to cut off from that frequency.

WATCH ONLINE *Element-2: 'Cutoff frequency keyboard note modulation'.*

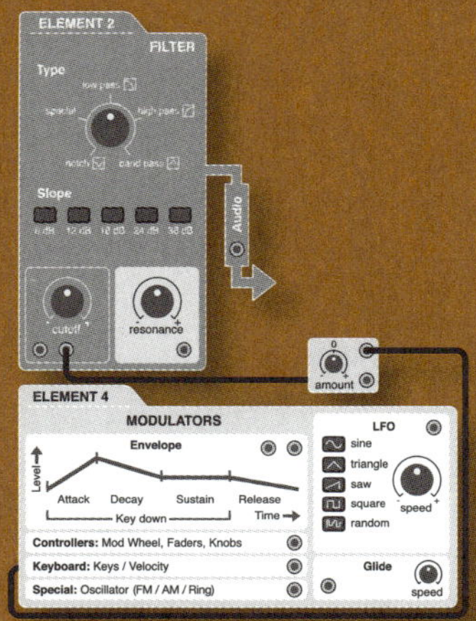


**Cutoff frequency keyboard velocity modulation**

In addition to note information being sent out by the keyboard, there is also velocity information. A typical keyboard has 128 steps of velocity values which go from value 0 up to value 127. It is wise to add an amount control between the keyboard and the cutoff frequency, otherwise the velocity control is going to be way too much.

The amount setting is very important and is basically the only control you have, unlike an LFO or an Envelope, which both have more parameters to tweak. The velocity of the keyboard is simply the value of how hard you play the keyboard. It's important that you set this parameter to your own taste inside the filter. The more amount you add, the more the cutoff changes and the less amount you add, the less the cutoff changes. So if you run through the presets of your own synthesizer, this parameter can influence expression within the filter. When you hear a preset and you don't hear your own expression in the filter, open up the amount.

### Cutoff frequency LFO modulation

LFO modulation is of course, related to Element-4. And now we already know what these Element-4 parts do, and that is to make other elements work and add dynamic changes. "Dynamic changes" is what we want to emphasize now. You will find more details about an LFO and exactly what it does in the Element-4 section of this booklet.

For now we're going to explain the LFO only in basic terms and how it functions in relation to the cutoff frequency. When you apply LFO modulation on the cutoff frequency you can hear a constant change within the sound. That is what an LFO does, it adds a constant repeating change at a certain speed. If I change the speed, the modulation is faster or slower. Cutoff frequency modulation by an LFO can be very nice for certain types of sounds. Especially when the LFO can be time-based musically, interesting sounds are possible.

*Element-2: 'Using LFO modulation'.*

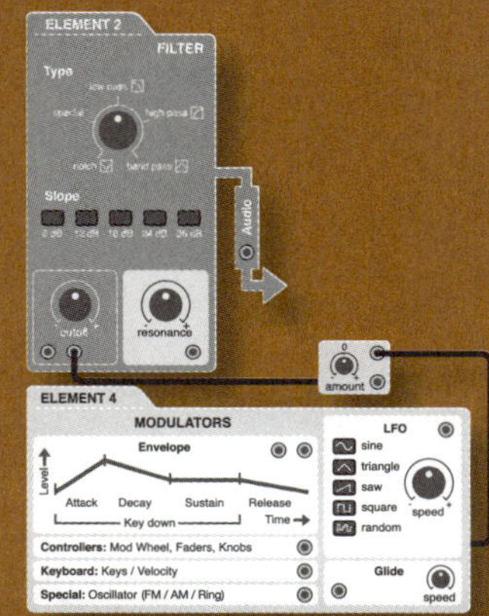

**Cutoff frequency controller modulation**

Two of the basic controllers on a synthesizer are the modulation wheel and the pitch wheel. But there are far more than just these, many synthesizers have freely assignable controllers, like the ribbon on the Alesis Andromeda or the beam controller on various Roland synths. A certain range in MIDI is reserved for all kinds of controllers. Because a controller changes other elements, it is inside the Element-4. Now, if we go back to the modulation wheel, it can be very useful as a controller applied to the cutoff frequency. Adding a controller to the cutoff frequency can be very useful to introduce more expression into the sound.

**Special cutoff frequency modulation**

The 4ES chart tells us that we can connect several parts into the cutoff frequency input. This means that if you have a synthesizer with enough options or a modular synthesizer, you can even connect an oscillator to it! So then the audio output of the oscillator will modulate the cutoff frequency. This is not used very often and is not considered a standard feature. The results are not always useful, but it is possible. The advantage of a modular system is that you can make all kinds of weird connections to try out. Some are not wise to do, mainly because they don't do anything, but it is good to experiment. The results are not always useful, but they can be very interesting in making special sounds or sound FX.

WATCH ONLINE *Element-2: 'Special cutoff modulation'.*

**Resonance modulation**

Please take the 4ES chart and look at Element-2. The resonance inside the filter also has an input. It has only one, but this is more symbolic than anything. This means that it is not commonly used in a modulation setup. It can be very interesting for changing sounds dynamically. For instance, you can modulate the resonance with an arpeggiator, or with a modulation wheel.

*Element-2: 'Resonance modulation'.*

# Element 3

# Element-3 / Amplifier

**Introduction**

*Introduction*

We are now at the final stop of the audio route of a subtractive synthesizer. It is called the amplifier and inside the 4ES we call it Element-3. As we said earlier in this training, Element-1, Element-2 and Element-3 are the audio parts of a subtractive synthesizer.

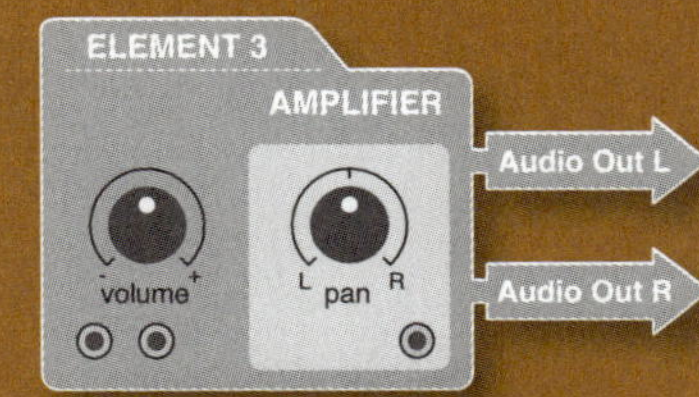


The audio signal starts at Element-1 and from there it goes into Element-2 and continues to the final stage, Element-3, which is the amplifier. We are already familiar with amplifiers from our studio setup or just think of a guitar player using a guitar amplifier.

But...why does an electronic device like a synthesizer have an amplifier and what does it amplify? Normally you amplify something that has no output in volume, right? A synthesizer has power of its own, so why does it have an amplifier?

Well, there are two reasons why. First of all, we need to amplify the signal that comes out of the filter. But most importantly we need control over the volume and this can only be done by using an amplifier. Look at the 4ES chart and you'll see that Element-3 has only two parameters, volume and panning. You won't find the panning control on all synthesizers and many will have volume as the only parameter of the Element-3/amplifier.

***Remember:***

- ***Volume parameter is the only amplifier parameter usually found on all subtractive synthesizers.***
- ***Panning is a parameter only found on more modern subtractive synthesizers.***

If we look at the 4ES chart we know right from the start what the real amplifier parts are. Just like the way we explained Element-2 and it's parameters....we're going to have some surprises in this topic as well.

The synthesizer panels we find in the real world are going to fool us again on the subject of the amplifier and...also on the subject of volume. More about that later on. First of all, let us show you the basic purpose of an amplifier.

***Element-3: 'Introduction Element-3'.***

As shown in the Video item we need an automatic control that controls the volume. The perfect tool to control the volume parameter is an envelope and once we connect that to the volume parameter it starts working all together.

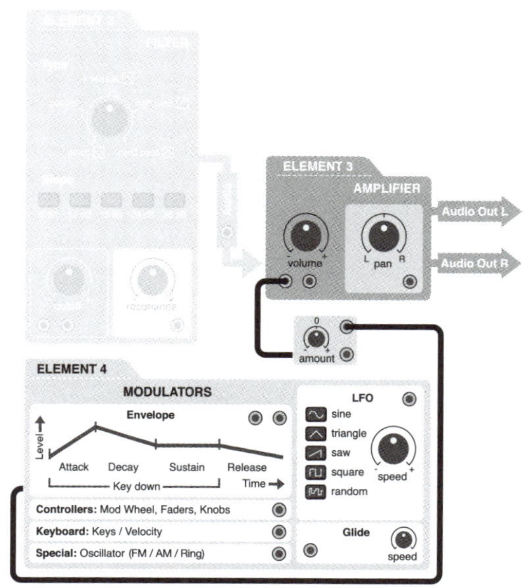

Once you hit a key, the envelope begins to work and controls the volume parameter. Basically without an envelope the amplifier is lost...and we'd be lost too, trying to use the synthesizer in a musical way.
Therefore, if we look at synthesizers in the real world, we can see that inside almost every amplifier section we're going to find an envelope that controls the volume.

As you know, the envelope is an Element-4 part and it's totally separated from the amplifier itself. Still...in general, we can say that an envelope connected to control the volume parameter is a basic need that allows the subtractive synthesizer to work in the way that it was intended.
One important note, this volume parameter is not the same as the volume knob on your synthesizer! We will discuss this in the next paragraph.

***Points to Remember:***

- ***Element-3 only alters the volume of the sound which comes from Element-2 and is called the amplifier.***
- ***Element-3 or the amplifier has only one major dynamic parameter called volume and sometimes an extra parameter called panning.***
- ***Element-3 has a basic need to have an Element-4 part called an envelope which controls the volume parameter.***
- ***Element-3 has the final word in the audio chain and incorrect settings on the envelope will cause faulty sounds.***

*Volume parameter of Element-3/amplifier*

### Volume parameter of Element-3/amplifier

In the introduction we explained to you that volume and panning are the only real amplifier parameters. All the other parameters you see on the amplifier panel section of a synthesizer are always Element-4 parts such as the envelope.

The envelope connected to the volume parameter is the main amplifier tool inside a subtractive synthesizer setup. But volume itself also functions as the overall volume setting for each sound. In this chapter, we're going to make a clear distinction between the volume parameter of the amplifier and the volume knob that you have on the panel of your synths. These are two entirely different things... and indeed it can lead to confusion.

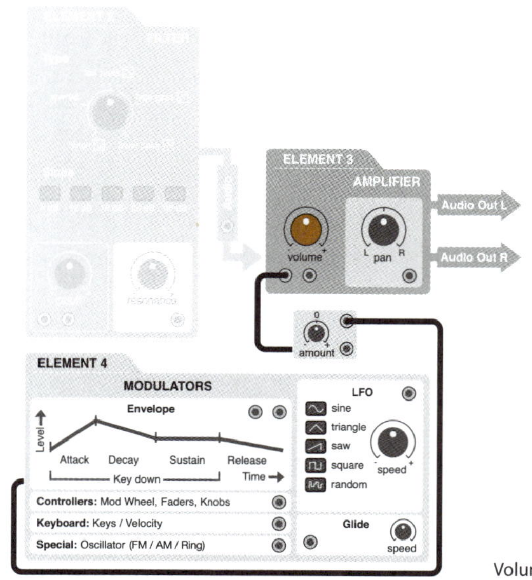


Volume inside amplifier open.

By itself, the volume parameter is the initial setting and if it's not controlled by an envelope, it's just a volume control for the signal that is coming from Element-2 which is the filter. When we open this initial volume parameter setting, we will hear that the sound continues all the time....which is not very musically useful.

Therefore we need to be able to close it and connect an envelope to it so we can control the volume parameter. The amount control regulates how deeply the envelope is going to affect the volume parameter. Only....this amount control is set up so that not only it controls the amount of the envelope, but it also functions as an overall volume control at the same time.

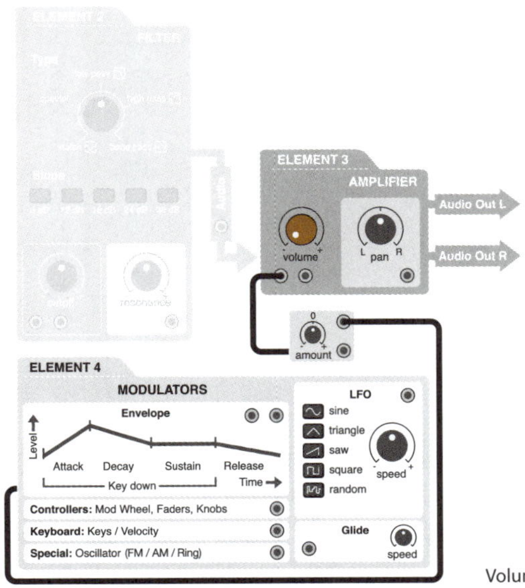


Volume inside amplifier closed.

This might be confusing to you, so let us explain it by taking another look at the 4ES chart. Inside the 4ES, we connect an amount control between the envelope and the volume parameter. This is the way it is done on all subtractive synthesizers. In order for us to have total control of the volume, we need to bring back the volume parameter itself to zero. Otherwise, we're going to get the same results as we did on the modular synth (Video item), where the sound kept on passing through and never stopped.

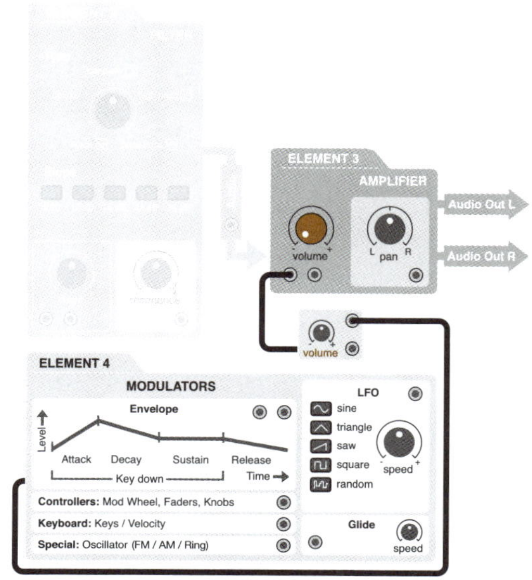


Envelope amount is displayed as "volume" inside synthesizers.

It seems odd that we close the volume parameter and open the envelope amount control, but you are actually will end up setting the volume parameter to zero....to make it all work! So the amplifier volume parameter inside the synthesizer is zero and the amount control that is in between the envelope and the volume parameter is the overall volume parameter. Take a look at the images to get a graphical impression of it. Sometimes, pictures can indeed be worth a 1000 words.

***Remember:***

- ***Volume parameter is the initial volume.***
- ***Once you open it, it adds a continuous bleeding of the signal that comes from the filter Element-2.***
- ***Envelope amount is the volume of the sound, all synthesizers call it volume instead of envelope amount.***

***Element-3: 'Volume parameter'.***

**Three types of volume**

*Three types of volume*

Now you may be wondering about the volume knob on your synth after studying the previous chapter on volume parameter. We've learned that it is basically the envelope amount and not the volume parameter itself that controls the actual volume of the synth. Because many synths also have mixers behind their signal path, and for convenience on stage and in a studio, there is actually a third volume control.

For instance on the left side of the Roland Jupiter-8 panel is a mixer part. Next to a volume knob you can also find a mix knob for controlling the lower and upper. With this you are able to combine two synth sounds on the Jupiter-8 . This is accomplished by using a split or dual mode.

The same thing goes for the master volume of the Virus b. Here you'll find a master volume knob that's separate from the synthesizer, because the Virus also has FX on board such as reverb, chorus and delays. So volume in the synthesizer section has a third and different meaning.

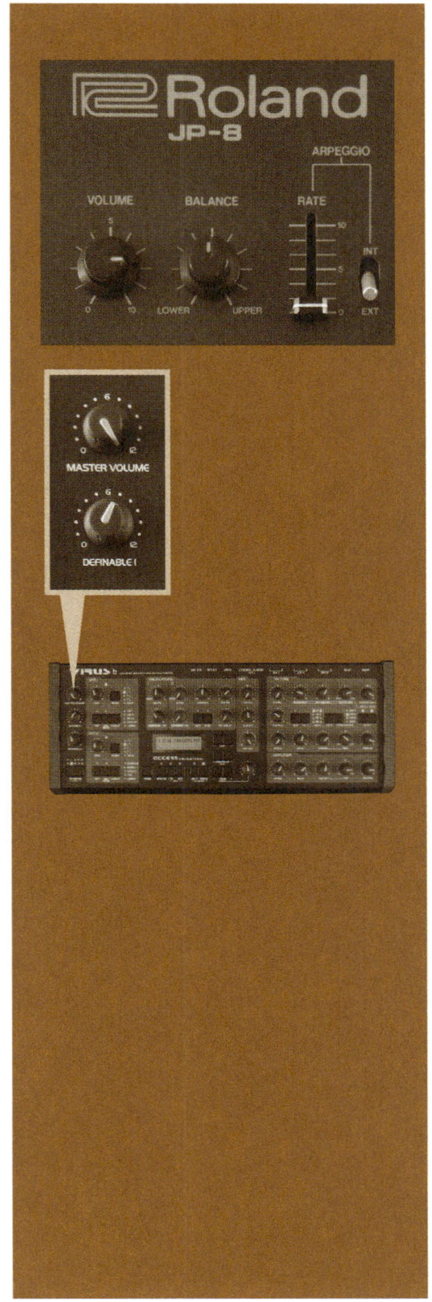

If we look at the 4ES chart, this type of volume, as mentioned in the Jupiter-8 and Virus b examples, is not depicted on the chart. This is because it is part of a mixer that's behind the output of the Element-3. There is one point you can remember that will help you to know which volume type you are working with. If you can change the volume of the sound, without affecting the preset volume, then you know you're dealing with mixer volume.

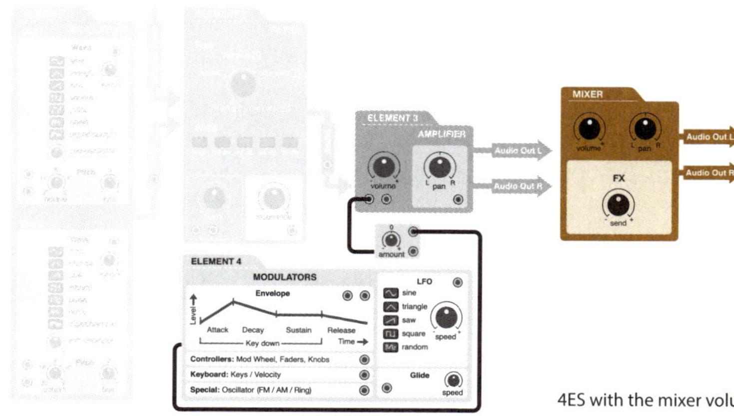


4ES with the mixer volume outside.

***Here's the three types of volumes:***
***1-volume parameter is the initial volume.***
***2-envelope amount is labeled as volume.***
***3-volume of the mixer inside the synthesizer.***

So the knobs labeled volume inside a synthesizer are:

- envelope amount control, being labeled as volume...because that is what it does in the end.
- mixer volume....that mixes the audio of the whole synthesizer including effects and stacked sounds.

***Element-3: 'Three types of volume'.***

## Amplifier envelope

*Amplifier envelope*

We've explained the volume parameter and how important the envelope is. Without it, we would be lost. Following that, we explained that the envelope amount control is in essence a volume knob to control the volume level of the sound.

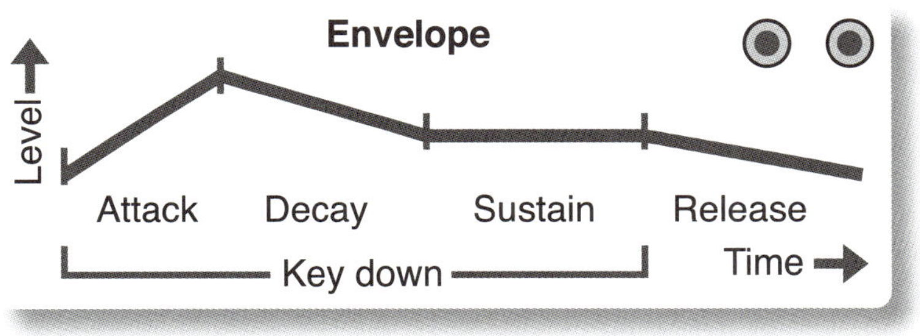


The settings inside this envelope are, of course, very important. It is the final control over the volume contour of a sound. A correct setting of the envelope parameters like attack, decay, sustain and release is very important. If you work on a bass sound and you have the attack parameter open too much you can turn the other knobs and parameters all day long...but you won't get a good bass sound. If you have the sustain parameter set incorrectly a pad sound won't be a pad sound.

The next chapter will be about Element-4 because the envelope is an Element-4 part. Inside that chapter we will go into more detail about what you can do with an amplifier envelope. A general statement about what an envelope does when it's connected to Element-3 and the volume parameter is that it shapes the volume contour of a sound, or the way the volume changes over time.

The designers of the MiniMoog mention this on the control panel of the MiniMoog, calling it the "loudness contour". If the setting is totally wrong...you won't get any sound. You can turn the other knobs endlessly but nothing is going to happen.

*Element-3: 'Amplifier envelope'.*

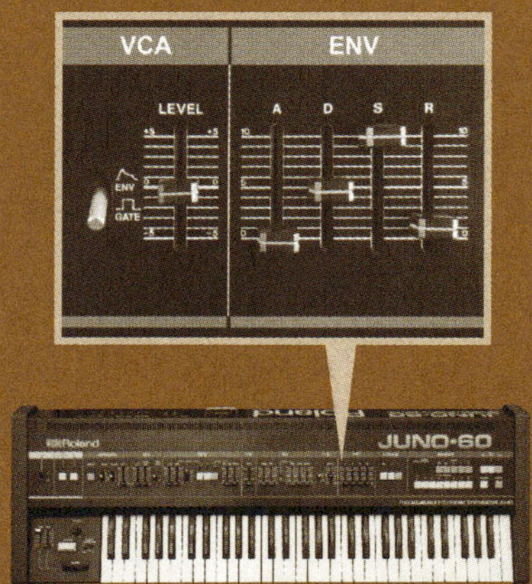


*Exception with no amplifier envelope used*

**Exceptions with no amplifier envelope used**

As is often the case in everything, there are always exceptions to the rules. This, of course, also happens when working with subtractive synthesizers. Some synthesizers have the option to use a gate rather than an envelope to control the volume parameter.

The Roland Juno-60 is set up this way. This term gate comes from the older modular synthesizers and it basically means 'note on' or 'note off'. The keyboard connection part inside a modular synth has a gate output. This gate is basically a voltage that is sent out once you press the key. The 'note on' is the 'gate on' and the 'note off' is the 'gate off'. Simple but effective: if you connect this gate directly to the volume parameter, what you have in fact, is an organ envelope. No special volume curves are happening using this arrangement. Sometimes a gate function to control the volume parameter comes in handy. But...it is very limited of course. If you set the envelope to a certain setting, it will do the same thing as a gate. All you have to do to create this effect is to close the attack, open the sustain fully and close the release fully.

***Element-3: 'Exceptions'.***

## Other volume parameter modulators

*Other volume parameter modulators*

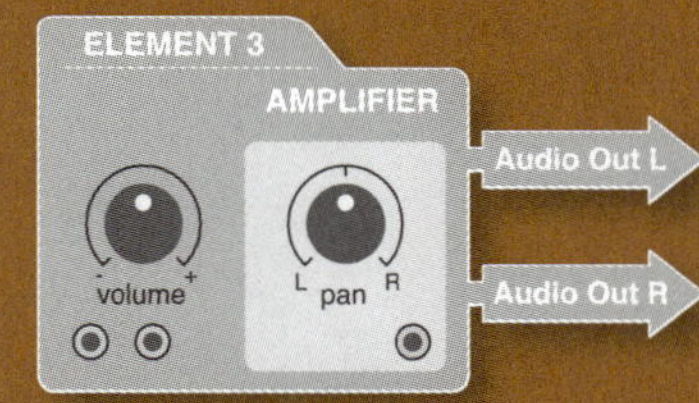


Keyboard velocity, LFO and many other parts from Element-4 can also control the volume parameter. We're going to discuss these in detail inside the Element-4 chapter.

The other parameter which you'll see implemented on many synthesizers is the panning control. By using panning in a static way you can set a sound anywhere in your stereo mix. But... there are other options available. What about stacking two sounds and pan one to the left side and the other one to the right side?

So, you see, this feature can already be very creative in a static way.
Additionally, we are able to modulate the panning parameter dynamically with other parts from Element-4, for instance a tempo-based sound.

As you see... we are again talking about Element-4 and we're going to get to that in the next chapter.

*In conclusion again remember:*

- ***Element-3 only alters the volume of the sound which comes from Element-2 and is called amplifier.***
- ***Element-3 or amplifier has only one major dynamic parameter called volume and sometimes panning as well.***
- ***Element-3 has a basic need for an Element-4 part called an envelope which controls the volume parameter.***
- ***Element-3 has the final word in audio and incorrect settings of the envelope will cause faulty sounds.***

*Element-3: 'Panning'.*

# Element

# 4

# Element-4 / Modulators

**Introduction**

*Introduction*

At the very beginning of this book we added a chapter called 'Before you move on'. It was essential to talk about Element-4 before we could explain the other elements. Element-4 is called 'modulators' and these are absolutely essential in producing sounds on a subtractive synthesizer.

A good illustration of what Element-4 is about and how essential it is to sound production, is to compare it to the human body and brain. Without the commands of our brain it's impossible to make our arms move or our legs....

Let us now reiterate the essential key to understanding Element-4. We also mentioned this in the chapter 'Before you move on'.

The key to Element-4 is: it makes the other elements work and adds dynamic changes.

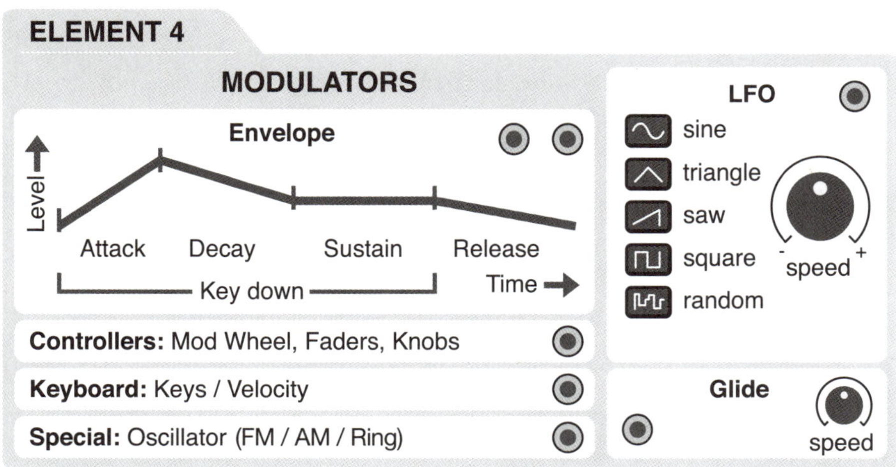


Figure 4.01

So all the movements and changes in sounds are controlled by Element-4! A very clear example that we saw in the chapter 'Before you move on', was the connection of the keyboard to the oscillators.

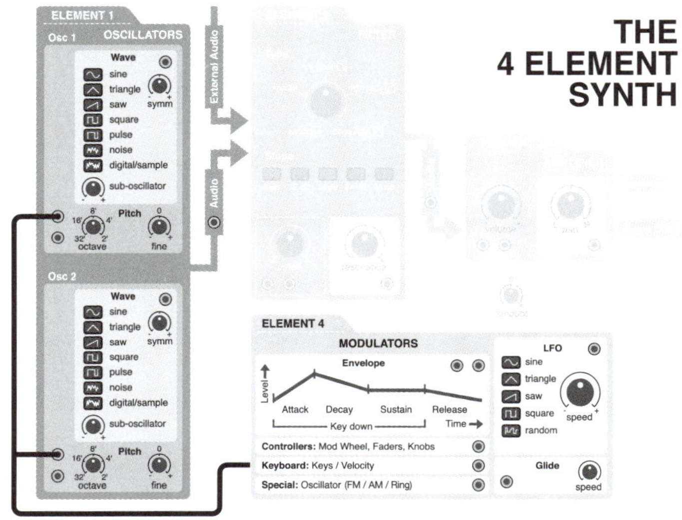


Figure 4.02

This is an essential connection because without it, we wouldn't be able to play any melodies on our synthesizers. This connection is a very clear example of what Element-4 is about, but there are also other connections that go from Element-4 into the other elements. Even connections from Element-4 back into Element-4 parts are possible.

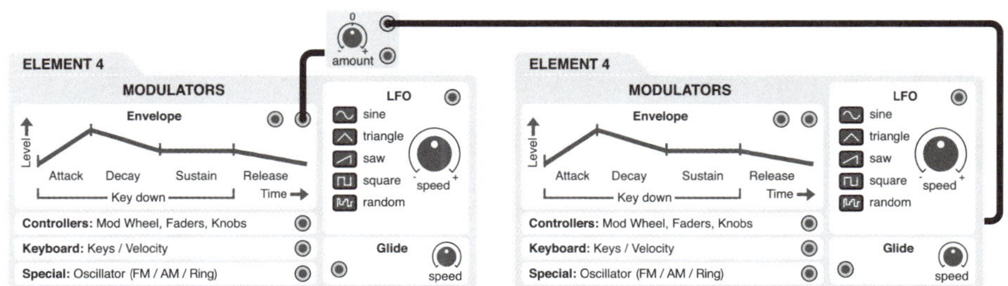


Figure 4.03

This is clearly illustrated on the 4ES chart. Using the 4ES chart will show us the answers to questions such as: "Can I connect this part into that part?" Just look at the 4ES and the answer is there. As long as a parameter has an input, we can connect other parts to it. Some connections will make more sense than others and some are essential, such as the connections we learned about in the chapter on Element-3.

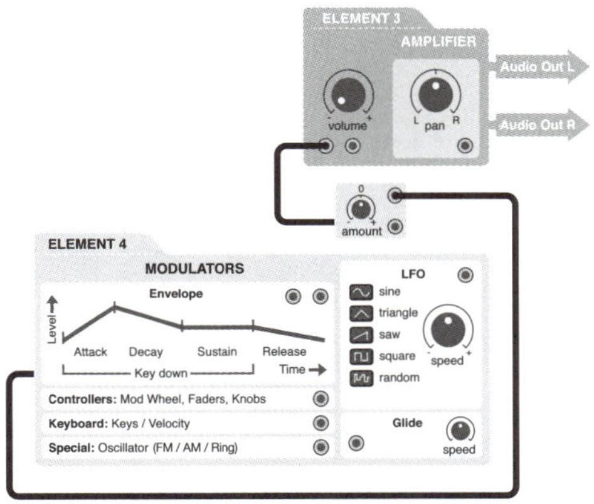


Figure 4.04

An essential Element-3 connection, for example, is to have an envelope connected so that we can control the volume parameter.

So what are modulators and what does modulate mean? In the yellow block called Element-4 we see the word modulators, but there are also independent parts inside of this block, such as an envelope, a LFO, controllers and a keyboard, and much more. What does to modulate mean in this case? Let's explain by using the Element-4 part keyboard. With the connection of the keyboard we can control the pitch of the oscillators. Usually we call this modulating, meaning that one part of an element changes the parameter of another element or even of the same element. In most cases this means that Element-4 parts are modulating the parameters of other elements. For example, if we connect an LFO from Element-4 to the pitch parameter of an oscillator it will also control the pitch, but in a very different way than when we use a the keyboard.

Figure 4.05

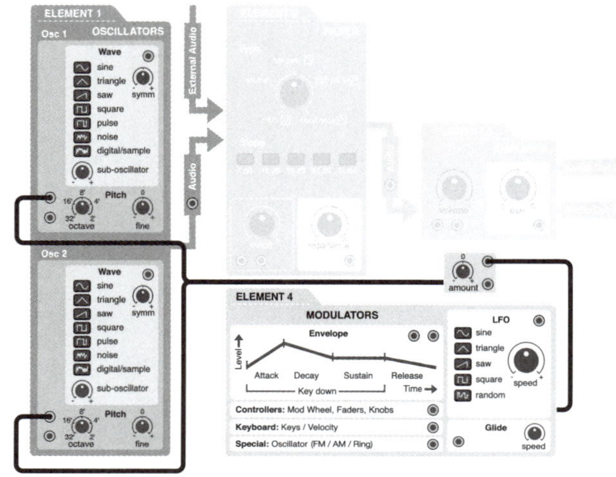


LFO connected to pitch for generating vibrato sound.

Figure 4.06

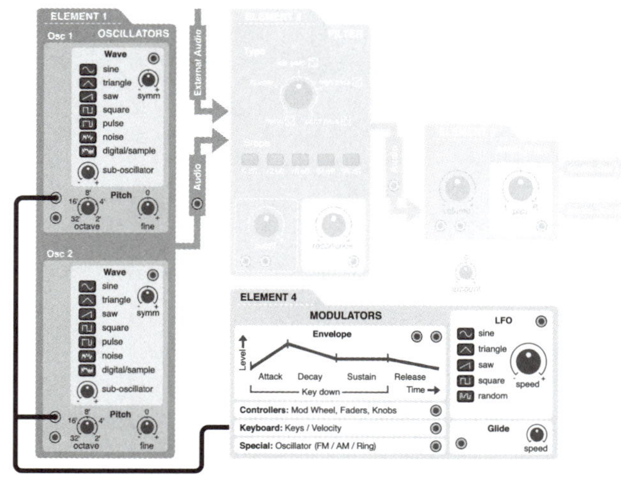


Keyboard connected to pitch to be able to play melodies.

The keyboard controls the pitch, so that we can play melodies, while the LFO adds vibrato to the sound. Both connections are modulating, but the outcome as it affects the actual sound is different. So with the LFO our natural response is to call it modulating and with the keyboard we say that it is controlling. But both of these connections use the same principles and we are still just connecting parts from Element-4 into parameters of other elements.

As you might have noticed, the 4ES chart is very important in allowing us to clear up any confusion that you might have about possible connections. All the parts that we find inside the Element-4 section of the 4ES chart are called modulators.

Also note that these parts are outside the audio route of the subtractive synthesizer. So no audio is generated here. We can also modulate Element-4 with parts from Element-4 itself. We'll learn more about this later on when we discuss Element-4 in greater detail.

Finally, we can say that modulating is simply connecting parts of a synthesizer to other parameters of that synthesizer.
In the following paragraphs we will use the term modulation very often, but we're sure you'll understand what it means by now.

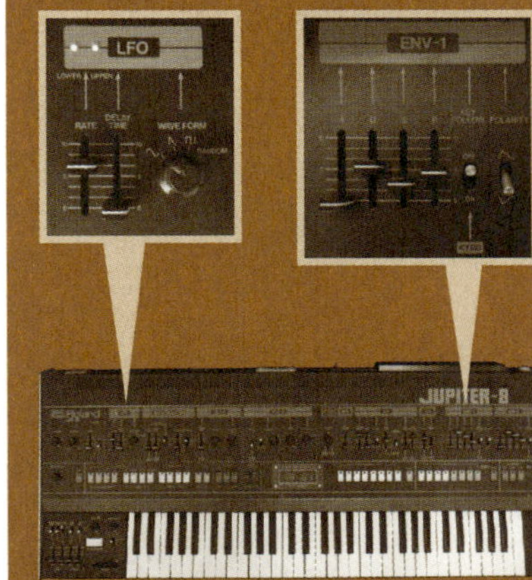


***Remember, the key to Element-4 is:***
***It makes the other elements work and adds dynamic changes.***

*Element-4 'Introduction Element-4'.*

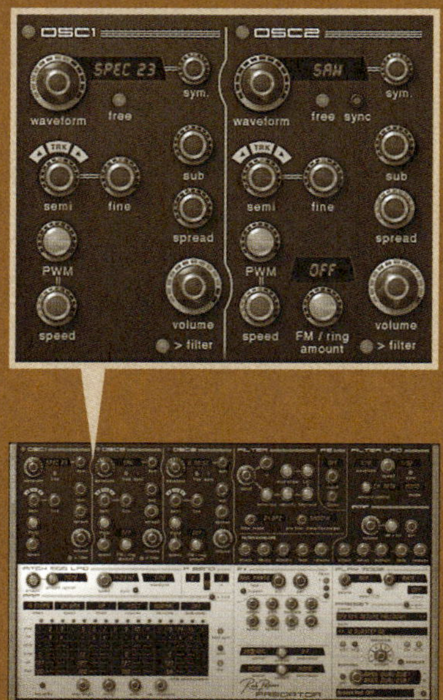


*Are all movements in sounds arranged by Element-4?*

**Are all movements in sounds arranged by Element-4?**

This is one thing you'll want to remember. We're going to run into the case where we hear changes in a sound.

Most are done with Element-4, but there is an exception to this. The exception is using a technique called detuning of the oscillators.

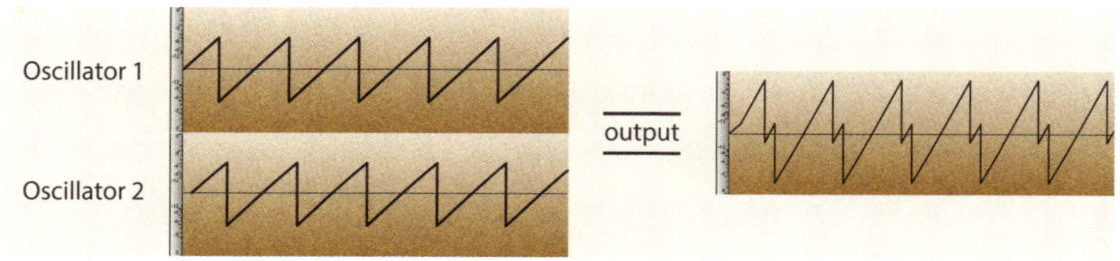


Once we use more than one oscillator, the oscillators mix together. This results in movement of the harmonics created by oscillator 1 and oscillator 2. The harmonics shift in and out of phase, and we can hear this movement. We can make the effect go slow and then we've created a kind of classic lead sound, or we can make it go fast and then we have a Dutch dance lead sound. But this movement is not controlled by any part of Element-4, instead, it's generated by the oscillators themselves.

WATCH ONLINE ***Element-4: 'Are all movements arranged by Element-4?'***

**Modulation amount**

*Modulation amount*

By now we know that the synthesizer is lost without the Element-4 parts and that without it nothing musical or useful can really happen. Now let's discuss common connections between the elements that are found on every synthesizer. We see on the 4ES chart a yellow box with a knob inside of it which is called amount.

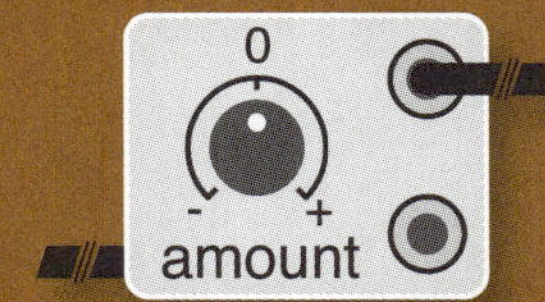


On the 4ES chart it is only one knob, but on a real synthesizer there's a lot of these. Let's explain why we need so many. For example, we've connected the keyboard directly to the pitch of the oscillators, so that we can play melodies on our synthesizer. But if we make this same direct connection with another part of Element-4, called an LFO, then we have a problem. The modulation we hear is far too deep. It's not useful, unless we want to make an effect sound. So we need something that will control the deepness or amount of this modulation.

**Figure 4.07**

Silver color knobs are the filter modulation amount knobs of Predator which modulate the cutoff frequency.

This we call an amount knob or amount controller. As we've seen before, on various synthesizer brands they use the word level instead of amount. Also a good term, of course, but most often the name used for this function is amount. In function, it is the same.

With an amount knob between the Element-4 part and the parameter it modulates, we can control how much modulation is applied to the parameter. This explains why we need so many of these amount knobs inside our synthesizer: so we can control the amount of modulation that is linked to a given parameter. With amount knobs, we can apply the modulation as much or as little as we want to create the sound that we want.

Maybe it is nice to compare this to using reverb on your mixer. We wouldn't usually add 100% reverb to a signal, but instead just a tiny bit by using the effect send.

Now we know what modulation is and that we need amount knobs to make it work in the right proportions. But do we hear all modulation? Well, that we're going to find out next.

WATCH ONLINE *Element-4: 'Modulation amount'.*

## Envelopes



The Element-4 part envelope is a prime feature of the synthesizer being used all the time. We saw this demonstrated in the Element-3 chapter, and also inside Element-2 where we looked at an envelope that modulated the cutoff frequency of the filter.

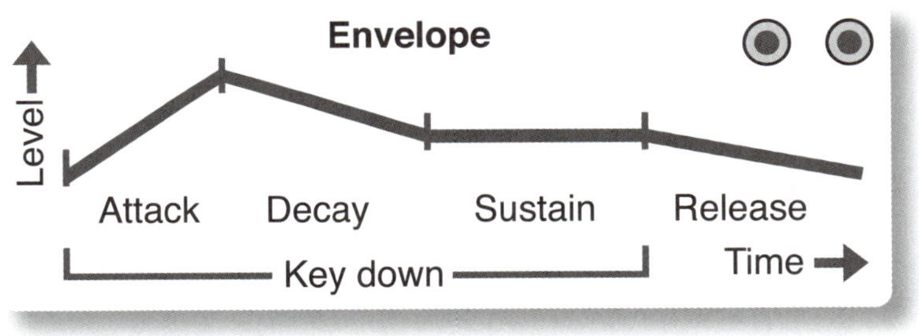


It reflects the essential key of Element-4: it makes the other elements work and adds dynamic changes. These dynamic changes refer very much to what an envelope does as an Element-4 part. The envelope adds changes over time to the parameter that we address it to. The clearest example of this fact is the envelope connected to the Amplifier. By using an envelope, we can build a sound with a slow attack, which is great for pad sounds. Or at the other hand, we can also create percussive synth sounds using an envelope.

The MiniMoog has a great name for the volume envelope. It's called loudness contour. This is a great name for explaining, because an envelope shapes the contour of the volume change. In any case, we're going to find envelopes on every synthesizer.

***Element-4: 'Envelopes / Introduction'.***

*How an envelope works*

## How an envelope works

The type of envelope we're going to explain in this book is a basic envelope. There are variations on this basic envelope, but we'll explain that later on. These are just variations of the basic concept. First, we need to connect something to the envelope to make it work and in most cases it is the keyboard.

Figure 4.08

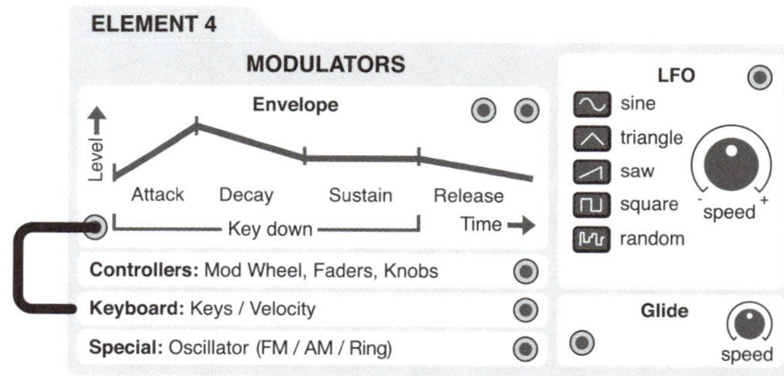


Keyboard connected to the envelope.

So, what we can do is connect the gate signal of the keyboard to the trigger input of an envelope. This is how we used to do this on a modular analog synthesizer. This connection is already pre-wired inside today's synthesizers. It's only on modular synthesizers that we'll need to connect the keyboard manually.

After being connected to the note on and note off information of the keyboard, an envelope will start to work.

An envelope is a time-based module with several parameters that can be used to control a parameter. The parameters of an envelope itself are attack, decay, sustain and release. These are the parameters of a basic envelope, which we're going to learn about in this book. The envelope makes a time-based movement from 0% to 100% level and then back again to 0%, and we control this behavior by using the four envelope parameters, attack, decay, sustain and release.

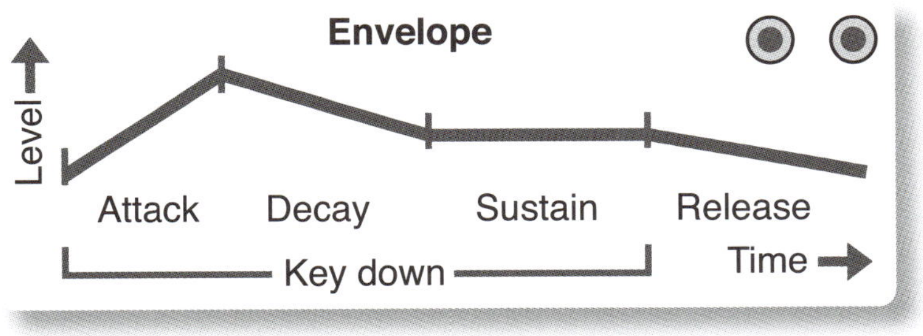

We'll want to keep the 0 to 100% movement of an envelope in mind. Because what the envelope does inside this range will depend on which parameter we assign the envelope to. So in the case of assigning the envelope to the volume parameter, the envelope moves from 0% to 100% volume and then back down to 0% volume again.

*Element-4: 'Envelopes / How an envelope works'.*

### Attack

Once we hit a key, the first parameter that is activated inside an envelope is attack. This is called the attack stage of an envelope. Attack determines how fast the envelope goes from 0% to 100% level. So in the case of volume, the attack parameter of the envelope determines how fast the envelope will go from 0% to 100% volume after hitting a key. The question is, how do we use the attack parameter effectively in creating our sounds?

Generally, using an envelope to control the volume is easy. For example, pad sounds use a slow attack, while we'd generally close the attack for bass sounds. The way you use the attack setting will vary depending upon the parameter we assign the envelope to control.

Overall we can say this about the attack stage:

- attack is the first parameter to act once we hit a key
- attack always goes from 0% level up to 100% level
- if attack is set to 0 (zero) time, the envelope starts at the 100% level from the moment we hit a key

Figure 4.09

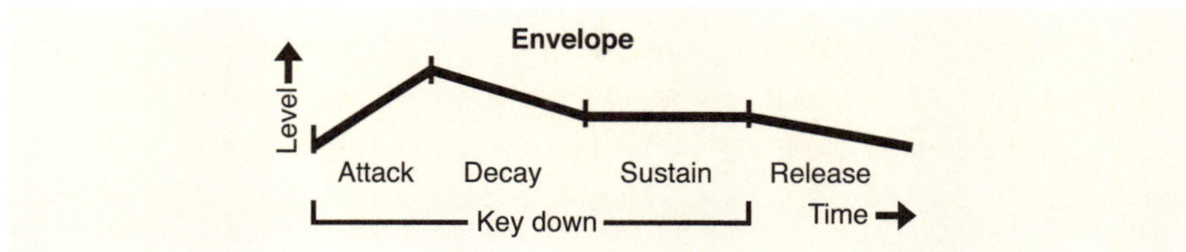


Slow attack time setting.

Figure 4.10

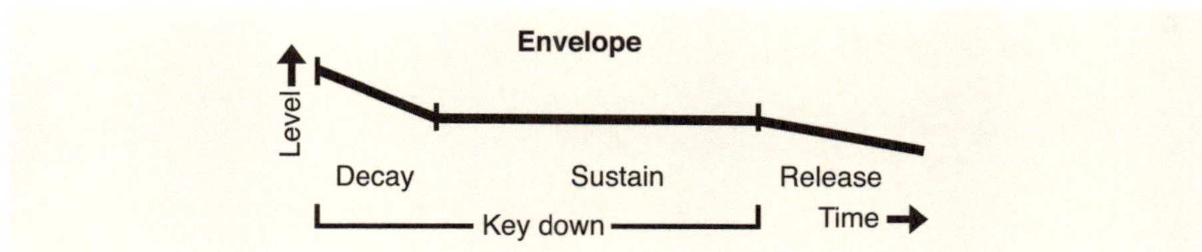


Attack at zero setting.

*Element-4: 'Envelopes / Attack'.*

## Decay

Once the attack time parameter has elapsed the decay stage is activated. So once the attack stage has reached its 100% level, the decay stage kicks in. Decay itself is the second stage of an envelope, but if we set the attack time to zero, then the decay stage is activated from the very first time we hit the key. The decay controls how much time it takes to go down from 100% level of the attack stage down to the sustain level. So when we use no attack time or in other words, set the attack time to zero, this means that the envelope starts right away from the 100% level, and so subsequently decay is the first parameter that changes the sound, going from 100% level down to the sustain level.

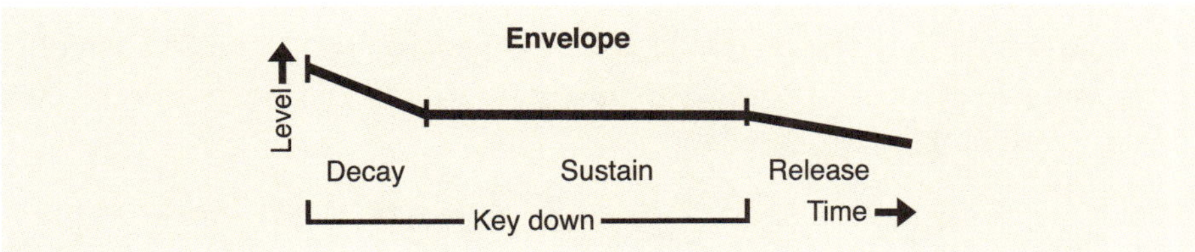


Figure 4.11

Sound from 100% level to sustain level.

The decay stage will always go to the sustain level. If we change the sustain level, the speed in which the decay goes down to the sustain level does not change. The sustain level can be 50% or 10% or 75% or whatever. This, of course affects the sound.

If we change the sustain to 100% level, then something special happens. It means that while the decay still can be changed, it doesn't affect the sound at all.

Figure 4.12

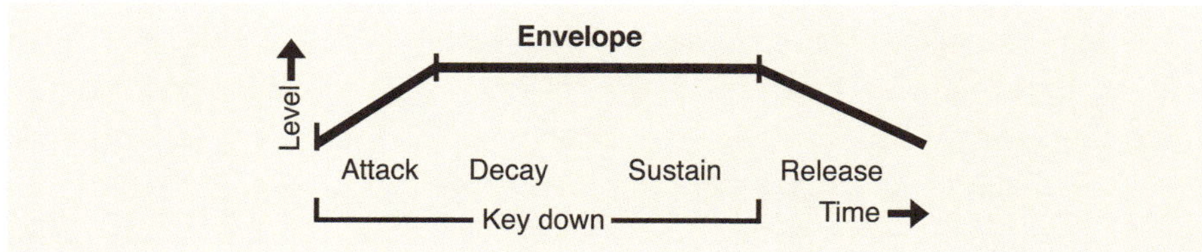


Sustain changed to 100% level, decay still can be changed but it doesn't affect the sound at all.

This is logical because the attack stage always ends at 100% level and then the decay stage takes over and goes on to the sustain level. But if sustain is set to 100% the decay has nothing to decay to!

So in conclusion: decay doesn't work if we have the sustain level set to 100%. On the other hand, if the sustain level is set to 0%, this means that the decay will go down to 0%. If an envelope is set to control Volume in this way, then the volume will go to 0%.

Overall we can say the following about the decay stage:

- decay starts to act after the attack stage reaches 100% level
- decay is the first parameter that starts to work if attack time is set to zero
- decay always moves on to the sustain level
- decay has no effect on the sound if we have the sustain level set to 100%

***Element-4: 'Envelopes / Decay'.***

**Sustain**

If we look at a modular synthesizer and its envelope, we can see that attack, decay and release usually have "time" written next to the knob.

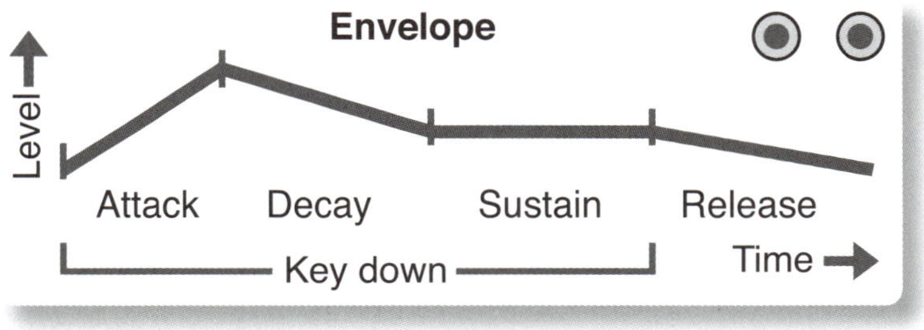

But we can also see that sustain has "level" written next to the knob. So sustain is not time based! Sustain is the level at which the envelope holds its level or in other words, where the envelope sustains its control signal. In the middle of the envelope, after the attack stage and decay stage, there is a kind of resting point. A place where we can sustain our envelope to a predefined percentage or level. In the case of an envelope that is assigned to the Volume parameter, the volume level will be sustained there. If an envelope is assigned to the pitch of an oscillator, then the sustain will hold the level of the pitch. What sustain will be holding really depends on what parameter we assign the envelope to work on. We'll explore this some more in the other envelope examples.

Overall we can say the following about the sustain stage:

- sustain begins to act after the decay stage ends
- sustain maintains the level of the envelope as long as we hold the key down

***Element-4: 'Envelopes / Sustain'.***

### Release

If we look carefully at the 4ES image, we can see that there is a bar below it which displays "key on" and "key off".

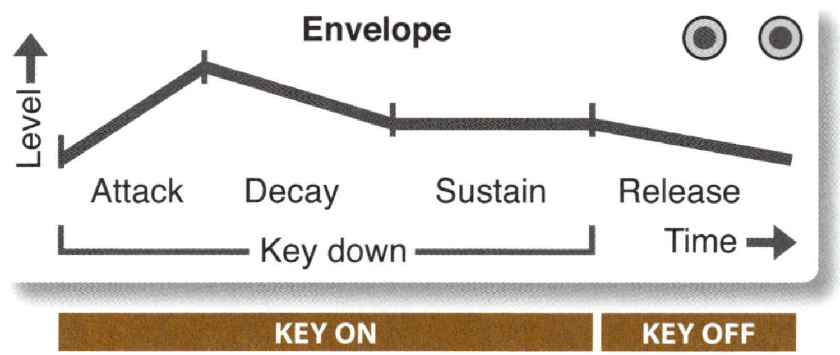


This is also known as "note on" and "note off" that we can find on many sequencers. What we can conclude after hearing the story about the sustain, is that these attack, decay and sustain level parameters work as long as we hold down the key. Once we release the key, the release stage starts to work. So the name of this parameter, release, is very well chosen: it happens only if we release the key.

The release stage is the last stage and makes the envelope go back down to 0% level. Of course we can change the release time and make it shorter or longer. But it will always go back to 0% level.

What happens if we change settings and set the sustain level to 0%? We're going to close the sustain and set the release to a certain time.

Figure 4.13

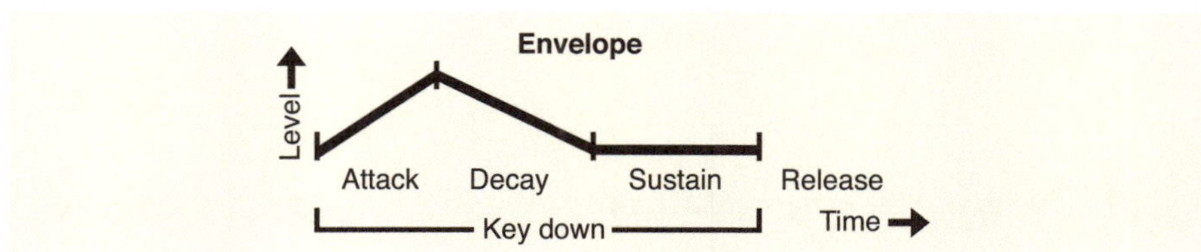


Sustain level set to 0%.

Also decay is set to a certain time. Remember we said that the envelope starts at 0% and goes to 100% level and then always returns to 0%. So this means that if we hold down the key, the decay works during the time we hold down the key. And if we hold the key down long enough, the envelope will go all the way back down to 0% level. And the reason for this is because our sustain is set to 0% level. The envelope is now at 0% level again, so if we release the key, the release stage doesn't work because the envelope has already reached its final 0% level.

If we release the key before the decay time has elapsed then the release stage takes over. For example we have a decay time of 5 seconds, but we let the key go after only 2 seconds. From that moment on, the release stage takes over and uses its own time setting to go back down to 0%

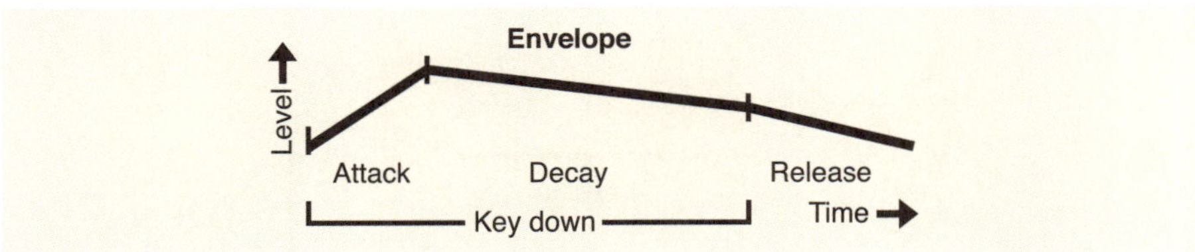


Figure 4.14

"Key off" during decay.

So when sustain is set to 0% level, we'll have to hold our key down long enough to let the decay stage finish to 0%. If we release the key more early the release stage will always take over because once we let go of the key, release starts to act as long as the envelope has not returned to 0% level.

Overall we can say these things about the release stage:

- release only starts after we let go of the key
- release always descends back to 0% level
- if an envelope has already reached 0% level, release will not be active

***Element-4: 'Envelopes / Release'.***

*Working with envelopes*

## Working with envelopes

We learned quite a bit using the example of an envelope assigned to the volume parameter. Is it a wise idea to connect the envelope to another part of the synthesizer? This changes the perspective but we're going to find out that the envelope still works from 0% to 100% level and back to 0% level again.

What's really going to change is its effect on the sound, this will be very different.

Figure 4.15

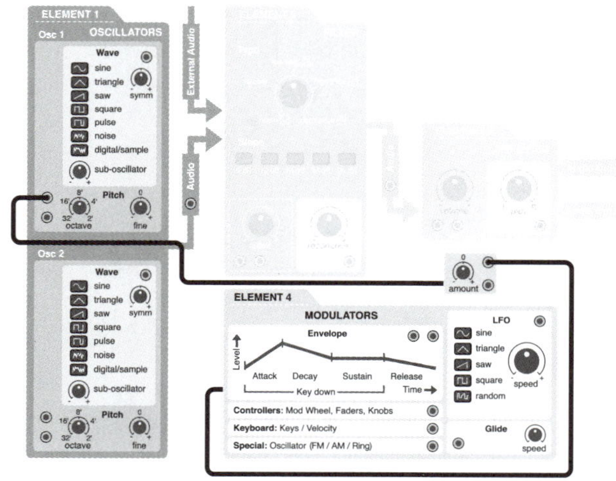


An envelope connected to the pitch of an oscillator.

For this example we connect an envelope to the pitch of an oscillator. Both volume and pitch are very clear examples of how envelopes function. But how about the filter envelope? This is an envelope that's connected to the cutoff frequency, and is found on every synth.

***Element-4: 'Envelopes / Working with envelopes'.***

## Three points to keep in mind using an envelope

*Three points to keep in mind using an envelope*

We've discussed the filter envelope already when we talked about Element-2 because it is an essential Element-4 part that we often find inside the panel section of a filter. We've already talked about the three points we have to keep in mind with the envelope that is modulating the cutoff frequency. Let's repeat it again to refresh our memories.

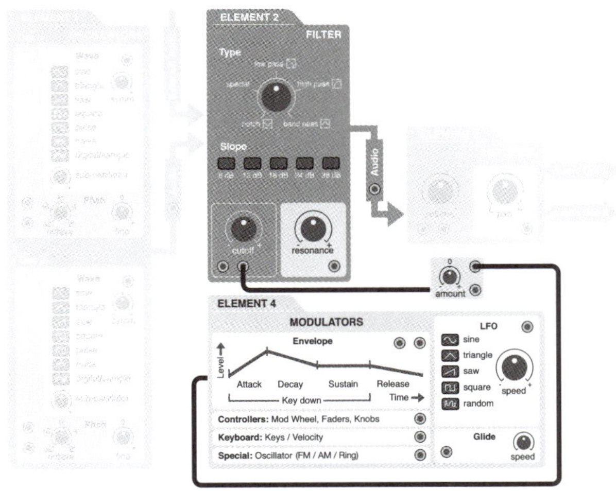


Figure 4.16

An envelope connected to the cutoff frequency.

If we look at the 4ES chart we can see the envelope with its parts attack, decay, sustain and release in the Element-4 section. These are the envelope controls.

The next step is to connect the envelope to the cutoff frequency. However, the connection of the envelope to the cutoff frequency is never a direct one, as there is always an amount knob in between the connection. The envelope is routed first into an amount knob before going on to the cutoff frequency. The amount knob is needed to control how much of the envelope we want to apply to our sound. This is because we need to apply a lot of envelope or a little, it all depends on the sound we are trying to create. We will find this connection on all synthesizers.

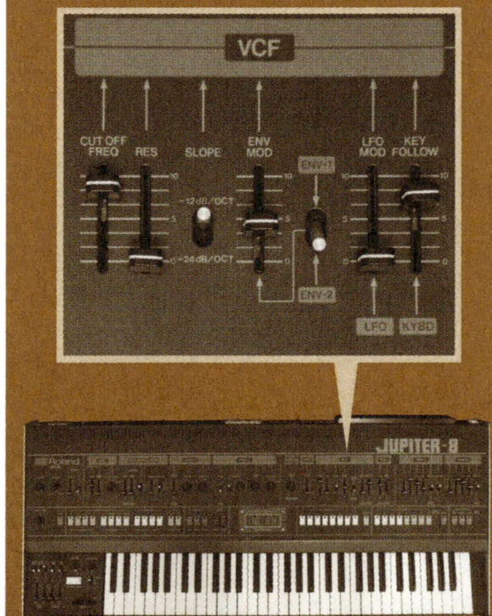


Now back to the envelope itself. Remember that an envelope with attack, decay, sustain and release is a time-based module inside a synthesizer. It is connected to the keyboard and once we hit the keyboard it starts to work. This is simply called cutoff frequency modulation by an envelope. By changing the settings of the envelope we can control how it affects the cutoff frequency over time. This means that we can create different sounds simply by editing the envelope parameters. We'll go into more detail, because there are a few things that we have to keep in mind when using the envelope to modulate the cutoff frequency.

There are three points of interest that are of great value if we're going to modulate the cutoff frequency with an envelope. These three points will give us endless sound options, so we need to know them to get the max out of our synthesizer.

**[1] Envelope and its settings**
On the panel of a synth we have attack, decay, sustain, sustain and release which are envelope parts, with their parameters and settings.

**[2] Cutoff frequency and its settings**
On the panel of a synth we can see the cutoff frequency as the central knob of the filter, but it is also the initial setting. In other words, it will set the initial point of the filter from where the modulation, in this case the envelope, will start.

**[3] Envelope amount**
On the panel of a synth we have an envelope amount knob and this controls how much the envelope is applied to the cutoff frequency. If this knob is closed the envelope will have no effect on the sound, but if we open it fully, the envelope will work at the maximum on the cutoff frequency.

***By keeping in mind these three points we now have a ton of sound options!***

Remember, there are also sounds that use no envelope modulation on the cutoff frequency, we can always leave it off. But on many sounds, the envelope is used to modulate the cutoff frequency.

We need to keep these three points in mind if we're going to use the envelope.

- cutoff frequency setting, this is the initial setting
- envelope amount setting, this sets how deeply the envelope modulates the cutoff frequency
- the settings of the envelope itself

This leads to a last important point about envelope modulation of the filter.

Many producers and synth players often only use the cutoff knob. They forget that there is also the envelope amount and the envelope settings that can change the sound. And not just by a little bit!

*Element-4: 'Envelopes / Envelope and cutoff frequency'.*

*Negative and positive modulation of an envelope*

**Negative and positive modulation of an envelope**

We explained what modulation means as well as the function of the amount knob, which we can find on the 4ES chart. The amount knob is needed to control the depth of the modulation.

If we look more closely at the filter envelope amount knob at some synthesizers and in particular Predator, we can see that it sometimes has a center position that says 0 or zero. So if we turn the knob left, we go into negative values! Until now, most of the time in this training we've shown positive modulation which is the most common. But negative modulation is also possible!

Figure 4.17

Predator filter envelope amount at zero setting, at positive setting and at negative setting.

This means that the way in which modulation works can be reversed. So, in case of our envelope, if we displayed it graphically the envelope would be upside down.

Figure 4.18

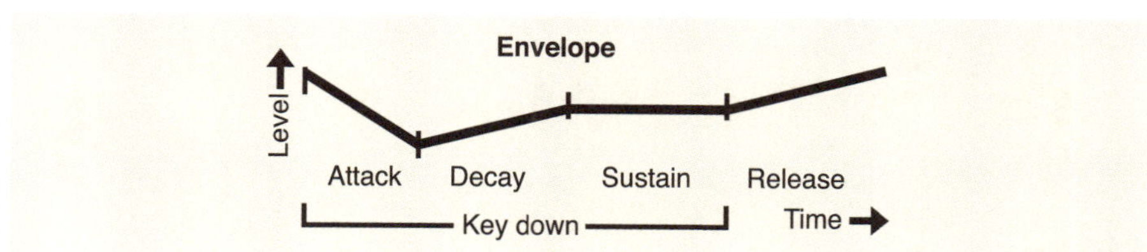


Negative modulation graphical displayed.

What we see happening with the negative modulation of the envelope is that attack becomes a descending stage, that decay becomes a rising stage, and release also becomes a rising stage.

Whenever this negative envelope is assigned to the cutoff frequency it offers us some different kinds of sounds. It's not used that often, but it's an interesting option to have for sure!

If we look at older synthesizers we can sometimes find a upside down envelope, like on the Jupiter-8.

It is labeled polarity which basically means that the modulation in the filter section is negative instead of the regular positive modulation. Most modern synthesizers offer positive and negative modulation, we can experiment with this feature if our synth allows it.

***Element-4: 'Envelopes / Negative and positive modulation'.***

## Other envelope parameters and variations

*Other envelope parameters and variations*

Until now we've explained the envelope in its most basic way of appearance as we'll find on many subtractive synthesizers. With the coming of the digital revolution, we saw new functions being added, which were sometimes just simply variations. But back in the analog age we already had delay options inside envelopes on some subtractive synthesizers. In general, delay, hold and sustain fade are by far the most used parameters next to the basic envelopes parameters of attack, decay, sustain and release.

### Envelope delay

The envelope delay is very easy to understand. It delays the start of the envelope. We can see this parameter inside BLUE.

Figure 4.19

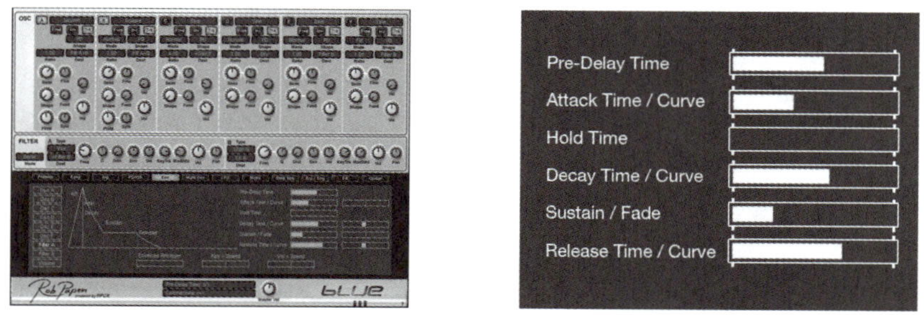


BLUE envelope delay.

The pre-delay time parameter as it is called is tempo-based. So once we open the slider, we can pick settings from 1/32T up to 16/1. A setting of 1/1 means that it will take 1 bar before the envelope begins to act. So the moment we hit the key, the start of the envelope will be delayed. Overall, we do not see this parameter implemented that often but it can be useful for effects sounds or for very special presets.

Figure 4.20

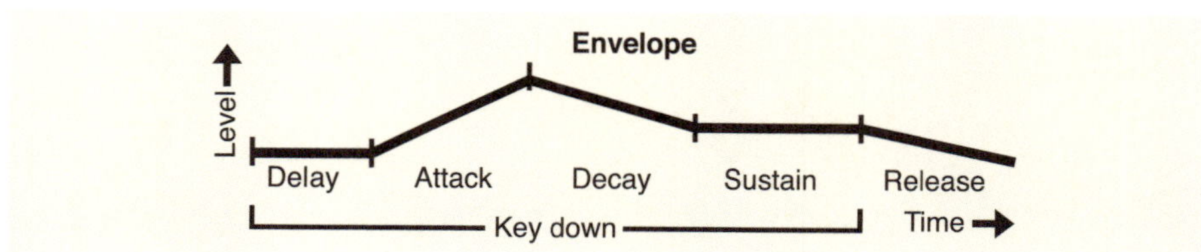


Envelope delay graphical displayed.

### Sustain fade

On some envelopes we have a parameter called sustain fade, or sustain time, we'll find this on several synthesizers. It is a very useful parameter, we see it more and more nowadays. Remember what we said in the sustain chapter, that sustain is level and it holds the level of the envelope as long as we hold down the key.

So once we have our sustain set to any level it stays there forever, as long as we hold the key down. By adding a fade time to the sustain we're adding a time factor to it. But where does the envelope go? With the attack stage we know it goes to 100% level and with the decay and release stages the envelope returns to 0% level.

But what level does the sustain time or sustain fade go to? We can see the answer on the panel of the Access Virus or Predator. On the Virus it says "fall" and "rise" and in the middle it has a symbol that means endless. In this case, endless means sustain in the regular way that we've already discussed.

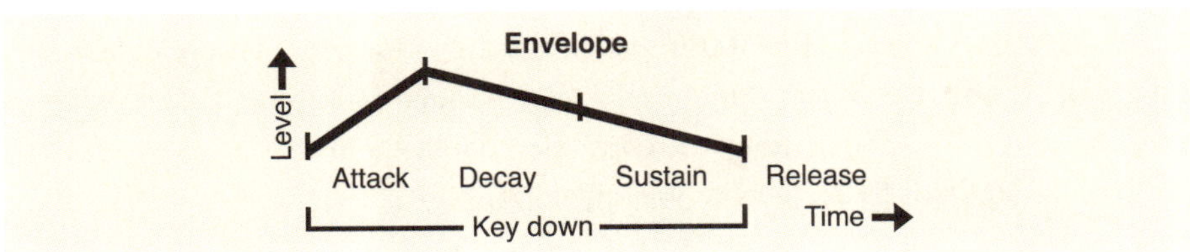


Figure 4.21

Negative or fall setting for sustain fade.

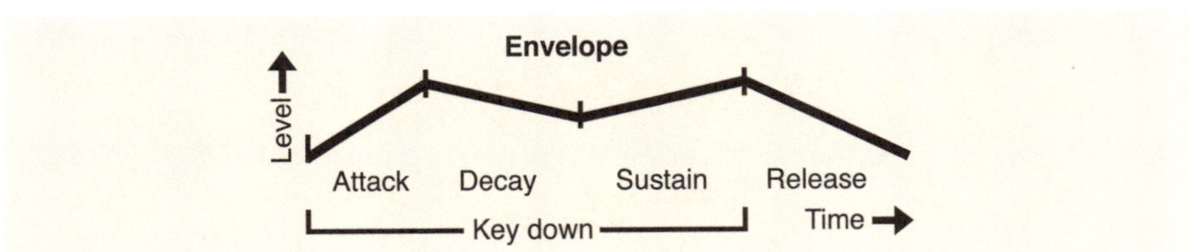


Figure 4.22

Positive or rise setting for sustain fade.

Figure 4.23

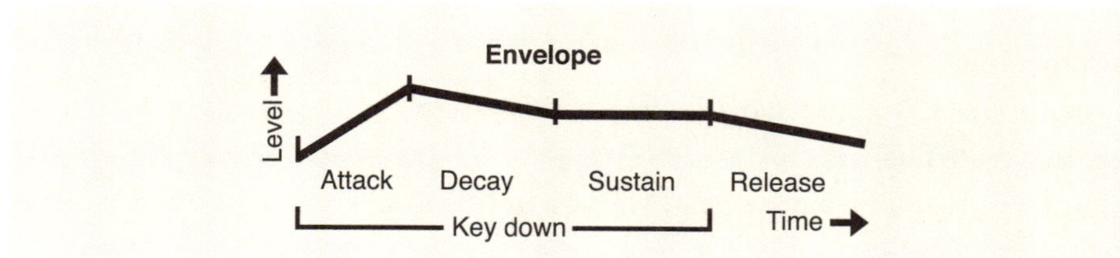


Envelope with sustain.

If we turn the time knob downwards towards fall this means that the sustain will be gradually lowered down to 0% level. If we move the knob upwards, past the endless symbol where it says rise, it means that the sustain will be gradually raised to the 100% level. So, depending on our settings the level goes down or up within a certain time. Note that our regular sustain stage has been cancelled out. So this is not going to be very useful for organ sounds or pad sounds.

Inside the filter section it's also very cool to use the sustain rise feature for brass sounds. What we hear is the filter cutoff frequency opening up if we hold the keys down long enough. It's a bit like real brass sounds with a lot of expression. Instead of keeping the cutoff frequency at the same level with the regular sustain, the sustain now rises again to the 100% level of the envelope modulation and creates a dynamic sound. Keep this in mind about the sustain fade/time: sustain fade or time changes the sustain with a time controlled parameter that moves upwards to 100% or downwards to 0% level of the envelope.

WATCH ONLINE ***Element-4: 'Envelopes / Sustain fade'.***

## Envelope hold

Envelope hold can be best explained as a second decay that maintains a 100% level. With decay we learned that it starts after the attack stage and it goes down to the sustain level. Envelope hold keeps the level the same. So it starts at 100% level after the attack stage, and it keeps it there. What is this good for then?

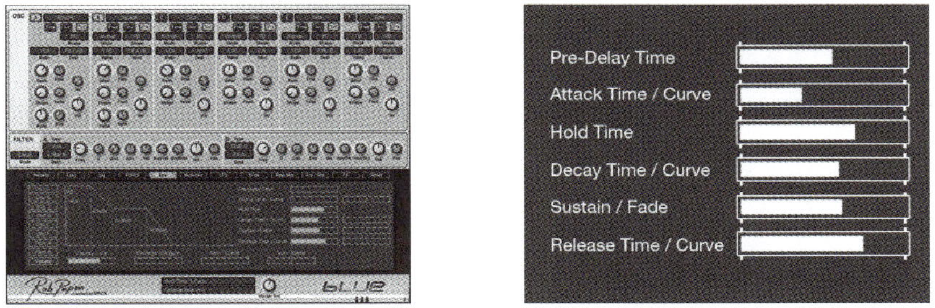


Figure 4.24

BLUE envelope hold.

For samplers this is handy because we can keep a sample, for instance a loop, playing for a certain time. Therefore, it's great for drums and percussion sounds. These types of sounds already have a natural decay to them and we do not want extra decay on a natural decay.

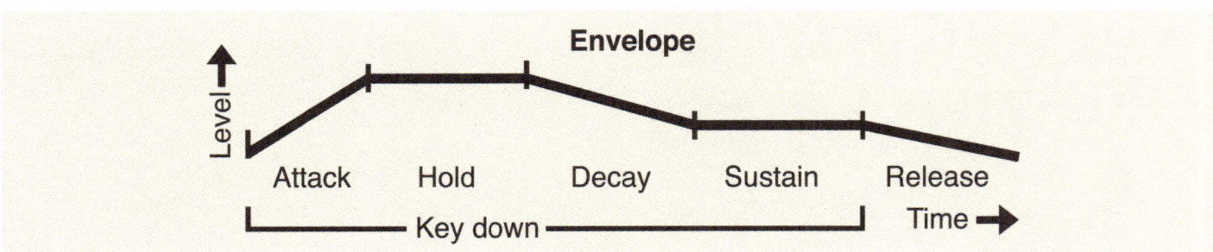


Figure 4.25

Key down untill sustain stage.

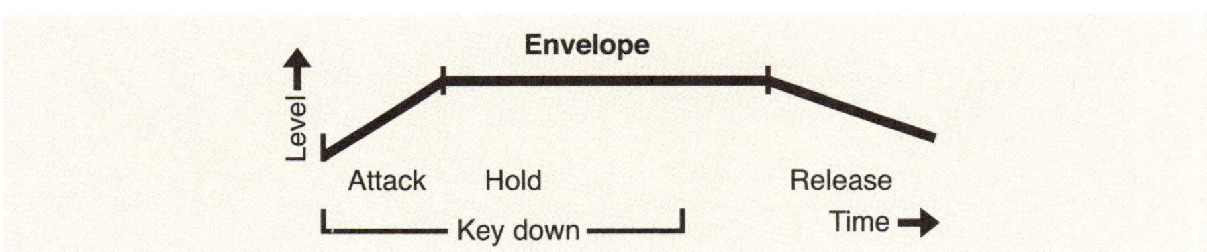


Figure 4.26

Key down untill hold stage.

Figure 4.27

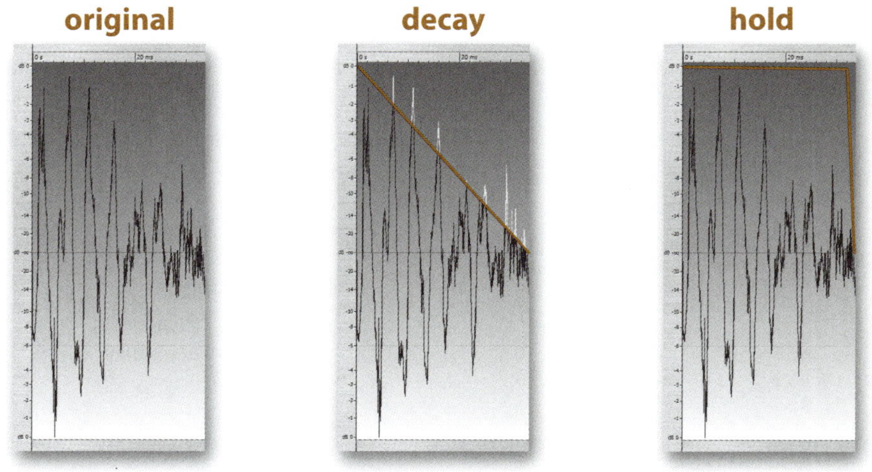


Snare drum sample.

Take a look at figure 4.22. On the left we see the original snare drum sample with natural decay. In the middle we see the same snare drum sample with volume decay added, results in altered sound. On the right side of the illustration we added volume hold, resulting in natural sound.

Envelope hold is perfect for this. Besides that, envelope hold is great for certain effect sounds: when it is used, a sound keeps playing and is less influenced by the duration of the key being held down.

### MIDI time-based envelopes

These are envelopes that have their settings of attack, decay and release expressed in musical bars and beats rather than milliseconds or seconds, like for instance in BLUE. These envelopes are great for tempo based presets and moving sounds. Of course, these envelopes are found mostly on digital software synthesizers. Still, it's a great feature that we can use to make some very interesting sounds.

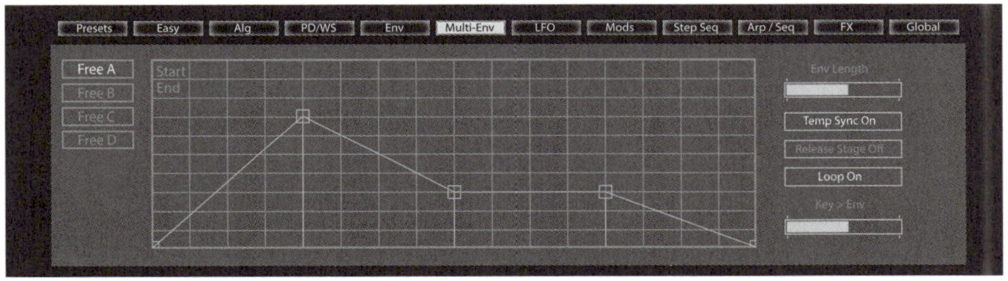


Figure 4.28

BLUE MIDI time-based envelope.

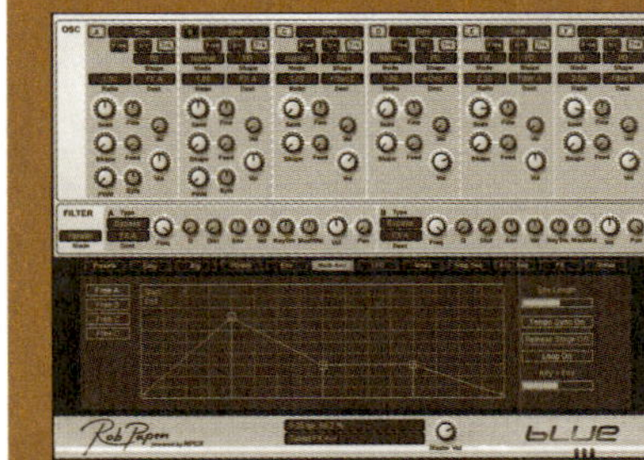

## Level stages

On many synthesizers we also have level as an option. So that attack appears as 'attack' and next to it is 'attack level'. Bear in mind this is just a variation on the basic envelope. An example of how this works is to set our attack level to 50%, and sustain set at 100%. Then decay will become a second attack stage because the initial attack stage did not reach the full 100% level.

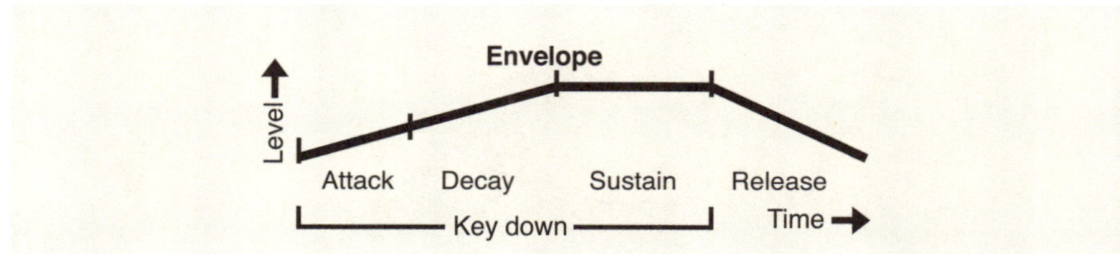


Figure 4.29 Envelope with level stages.

On the other hand, if we keep our sustain level below 50%, then this attack level is nothing more than an amount control, since the envelope will not use its full 100% range. In this specific case we could also set the modulation amount to 50% of the full envelope to accomplish the same thing, so this extra attack level tends to do nothing but cause confusion.

The extra control that is added by having this level function comes with a price. The level stages give us a bit more freedom, but they really make having overview harder.

To keep life easy, keep levels standard, and if you need to use these level parameters on your synth, read the manual of your instrument very carefully!

### Other envelopes

Well, there are certainly more types of envelopes that we can explore. Like the ones we can have inside BLUE. There's multiple stage envelopes and all kinds of wicked features. But again, these are just variations on the basic envelope. Once we understand the basic principles of an envelope, the rest is a matter of reading the manual and using trial and error to create the sounds we desire.

For instance, in BLUE a sound can have the repeat switch of the envelopes set to on. This makes the envelopes repeat according to musical bars and beats setting and will allow us to make exciting, moving new textures. Still, like we said before, just variations on the original basic envelope.

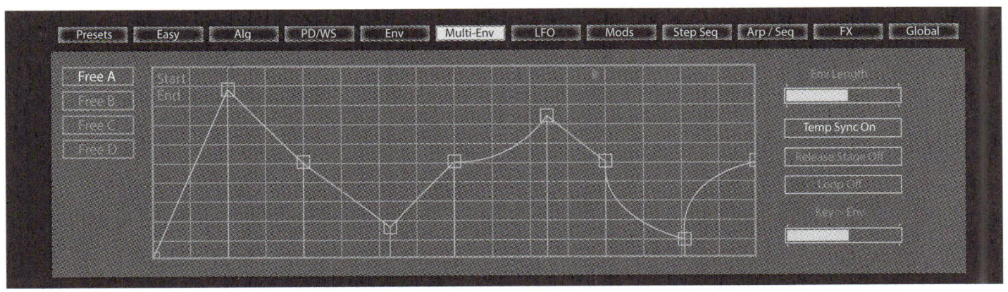


Figure 4.30

BLUE multiple stage envelope.

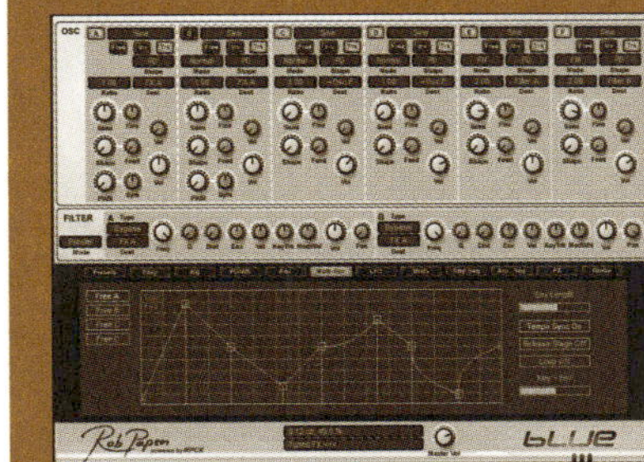

*Envelope modulation and Element 1*

## Envelope modulation and Element-1

Let's have a look at the 4ES chart, the basis for this entire training course. By adding an amount control to the envelope we see inside Element-4, we can direct it towards another parameter and have the option to add just a little of modulation or a lot of modulation. The options inside Element-1 and the parameters that we are able to modulate are also revealed by looking at the 4ES chart.

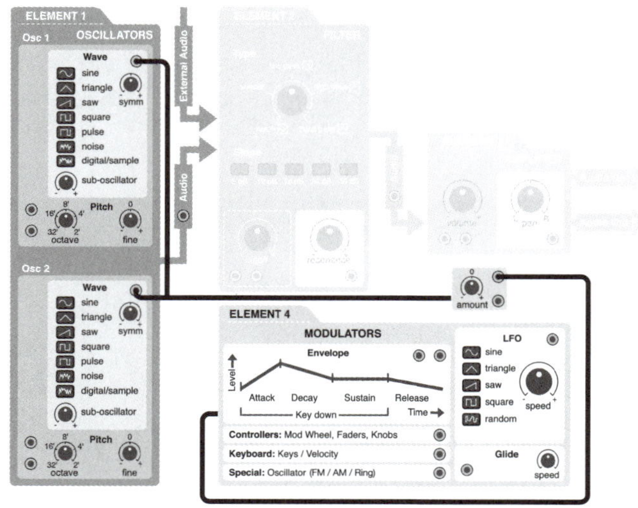


Figure 4.31

Element-1 and Element-4.

We can see two basic parts inside each oscillator, a wave part and a pitch part. Both have been given a slightly different color. The pitch of the sound is generated by the oscillator and the color of the sound is the wave that we use in the oscillator. Well, we can modulate both of these parts. We can conclude this from looking at the 4ES chart and seeing that there are inputs inside both the wave and pitch parts.

***Element-4: 'Envelope and Element-1' pitch'.***

### Pitch modulation by an envelope

We've already used an envelope on the pitch part to explain how an envelope works a few chapters back. Because of this pitch change the workings of an envelope are quite noticeable. The envelope connected to the pitch is great for making pitch sweeps. This can be used for FX sounds, but also for brass sounds, trance sounds, sync mode sounds, percussion and drum sounds.

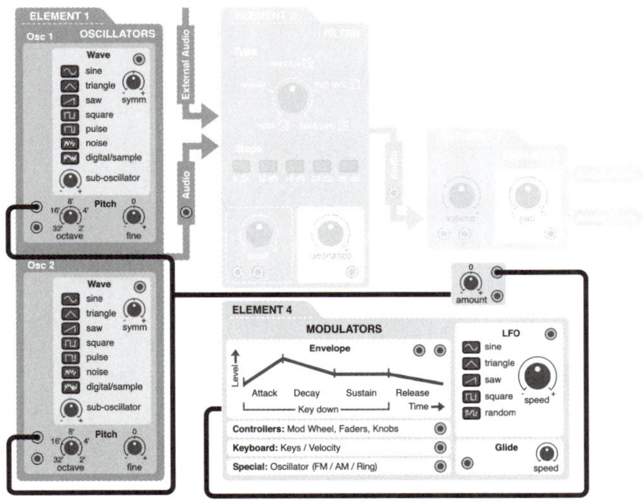


Figure 4.32

Element-1 and Element-4 pitch modulation.

### Waveform modulation by an envelope

We explained the modulation of a waveform in the chapter on Element-1. If we look again at the 4ES chart, we can see that only one input is shown. This shows us that waveform modulation is not used as often as the modulation inputs of the pitch of an oscillator. Of course we can connect more parts into the wave modulation input. The single input is only symbolic, just to display that it is not used very much.

Figure 4.33

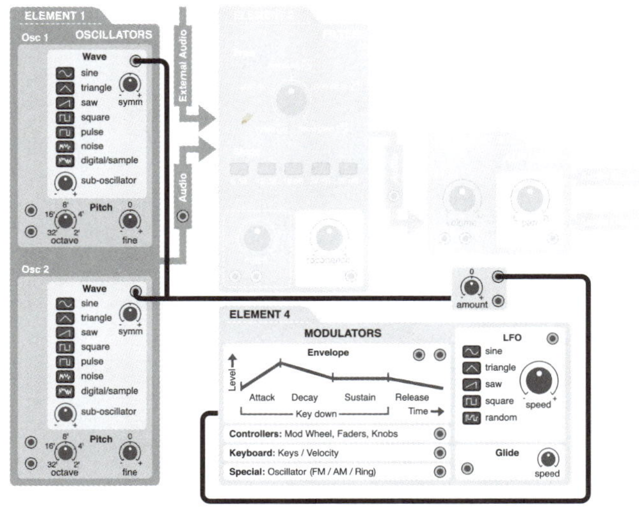


If we connect an envelope to this input, we modulate the symmetry. Like we said in the Element-1 chapter, this symmetry modulation has the best results when used on the square/pulse waveform. It changes the width of the pulse waveform.

Figure 4.34

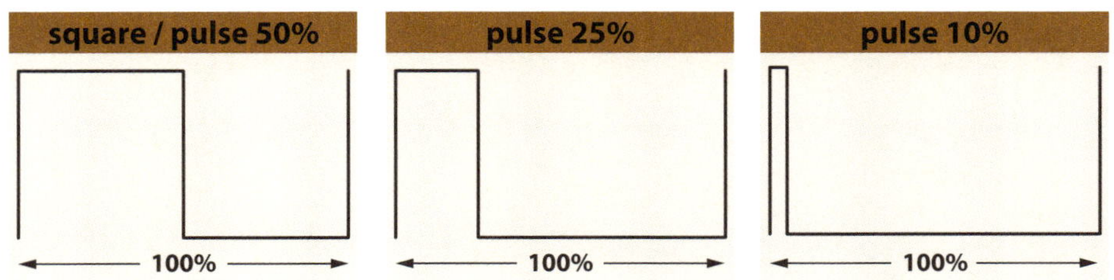


Square/pulse waveform.

On older synthesizers we won't find the name symmetry very often, because this modulation was used only for the Square/ Pulse.

This was done simply because the effect is minimal if the symmetry of for instance a Saw wave was modulated. Now with our modern digital synthesizers it is interesting to have the option to modulate the symmetry of other waveforms. The results are not always good, but it's interesting to give it a try. Using the envelope to modulate the symmetry can add interesting new textures.

*Element-4: 'Envelope and Element-1 waveform'.*

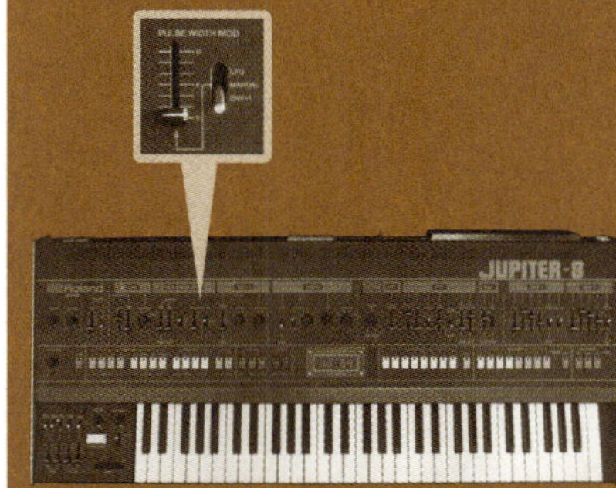

*Envelope mudulation and Element-2*

## Envelope modulation and Element-2

In the introduction to cutoff frequency modulation and also in the chapter about Element-2 we explained that most knobs on a filter panel section are used to modulate the cutoff frequency. The cutoff knob, resonance knob, and, if available, the filter types are the only real filter parameters.

The cutoff frequency knob is not only the central and main parameter of the filter but it is also heavily modulated by the envelope. Therefore we often find envelope parameters inside the panel section of the filter. Like on the Mini-Moog, on the Virus, and on the Predator and so on.

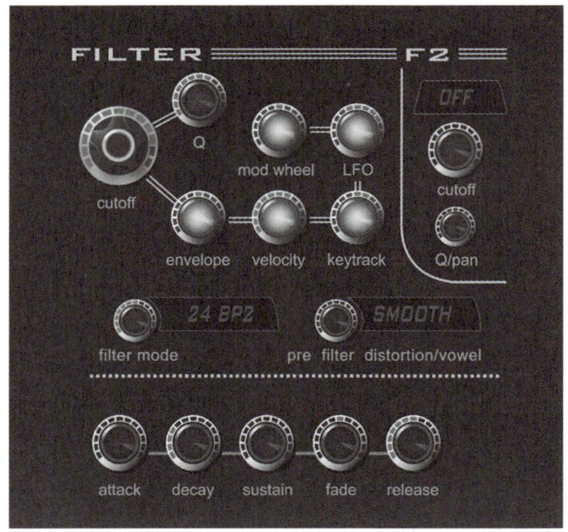


Figure 4.35

Predator's envelope parameters.

The envelope can control the cutoff frequency and with the amount setting we can adjust how much affect the envelope will have on our sound. Next to that, the settings of the envelope parameters themselves are essential. In fact, as we explained in the Element-2 chapter, there are three points that have great value if we're going to modulate the cutoff frequency with an envelope. These three points will give us endless options in sound creation, so we need to know them to get the max out of our synthesizer.

These three points are:

**[1] Envelope and its settings**

On the panel of a synth we have attack, decay, sustain, sustain and release which are envelope parts, with their parameters and settings.

**[2] Cutoff frequency and its settings**

On the panel of a synth we can see the cutoff frequency as the central knob of the filter, but it is also the initial setting. In other words, it will set the initial point of the filter from where the modulation, in this case the envelope, will start.

**[3] Envelope amount**

On the panel of a synth we have an envelope amount knob and this controls how much the envelope is applied to the cutoff frequency. If this knob is closed the envelope will have no effect on the sound, but if we open it fully, the envelope will work at the maximum on the cutoff frequency. Imagine how many sounds we can make using just these three points and modulate the filter. So, it's important to realize that all these parameters are influencing the setting of the cutoff frequency at the same time. We have to keep track of all three, otherwise we'll lose control. One thing is for sure, we'll find a dedicated filter envelope on almost all subtractive synthesizers. There is only one exception to this rule: the modular synthesizer. Here we have to connect one of the envelopes to the filter manually.

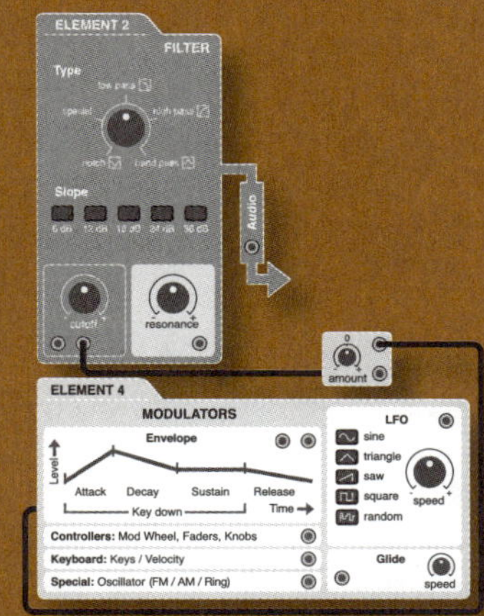

Now let's focus on the envelope parameters and their workings in relation to the cutoff frequency.

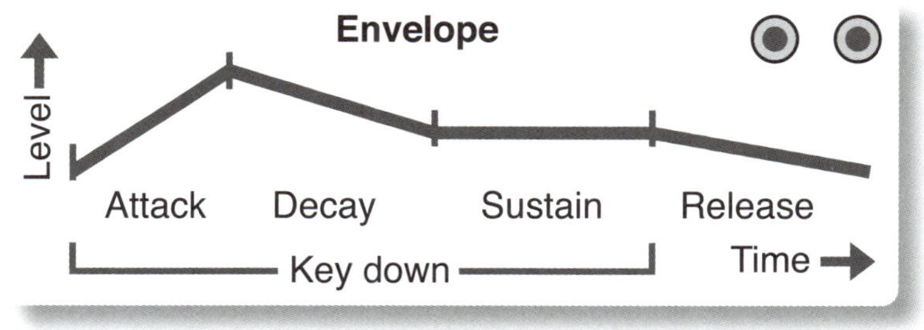


**The attack stage**

- attack is the first parameter that will begin to act once we hit a key
- attack always goes from 0% up to 100% level
- if attack is set to 0 (zero) time, the envelope will start at 100% level from the very beginning

The envelope is connected to the cutoff frequency and this means that the level of the envelope controls the cutoff frequency. So by using a slow attack time we make the filter open slowly. This is great for filter sweeps and brass sounds. Setting the attack time to 0 (zero) is also great for percussive sounds, bass and many other sounds that don't use a slow attack. The envelope starts at the 100% level from the moment we hit a key.

**The decay stage**

- decay starts to act after the attack stage reaches the 100% level
- decay is the first parameter that starts to work if the attack time is set to 0 (zero)
- decay always moves to the sustain level
- decay does nothing if we have the sustain level set to 100%

The decay parameter on a filter envelope is very important and allows us to go from one type of bass sound to another type of bass sound. The decay stage moves to the sustain level which is the next stage. If we tweak both decay and sustain a bit, we can hear that a lot of sound options present themselves.

**The sustain stage**

- sustain starts to act after the decay stage ends
- sustain holds the level of the envelope as long as we hold the key

So if we change the sustain of the filter envelope, we can hear very clearly the effect it has on the sound and how important sustain is for certain types of sounds. All kinds of different bass sounds are possible simply by changing the sustain level, and if we go a step further by also changing the decay stage ahead of it.

**The release stage**

- release only starts after we let go of the key
- release is always going to descend to 0% level
- if an envelope has already reached 0% level, release is not active

The envelope controls the cutoff frequency so in this case, release brings the cutoff frequency down to its offset level. Keep in mind with filter release that there's also the release parameter on the volume envelope and it will be active at the same time we press the key. So keep the release of the filter envelope in line with the release of the volume envelope.

We've already explained how an envelope works in general, but as we can see, inside the filter the envelope plays a very important role. In fact, without it the synthesizer would be very limited in the types of sounds it could create.

One trick to give the envelope on the filter some more bite in the 'contour' it produces, is to increase the resonance parameter of the filter. This makes a sweep sound sweep a lot more. If we cut the resonance back we will hear that the sweep is less prominent in the sound. Of course, this depends on the type of sound we're seeking to create. But the resonance parameter is a great tool to emphasize the movement of the filter envelope.

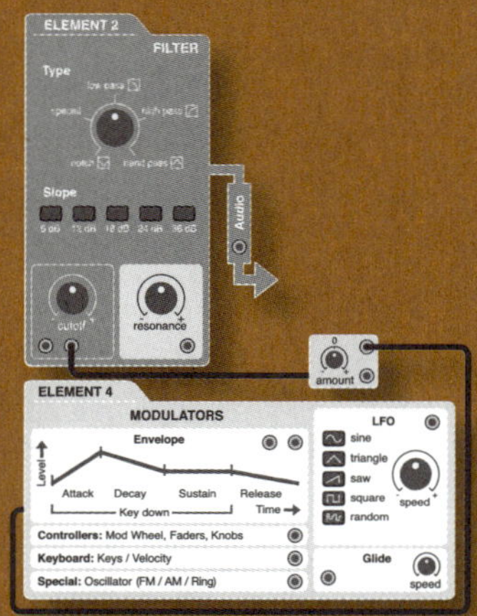


In subtractive synthesis, the filter and the envelope are essential. The basis of a sound starts with Element-1 and if the sound is wrong there we can change the filter and its envelope endlessly, but we'll never get the sound right. If we started with a sinus waveform inside the oscillator and then tried to build a brass sound, we'll never get there. So it all begins with Element-1 and the oscillators.

After that, the next important step is the settings of the filter and the filter envelope. Countless sound options are there if we take the three points to heart that we mentioned before in using the envelope, envelope amount and the cutoff frequency.

WATCH ONLINE ***Element-4: 'Envelope and Element-2'.***

## Envelope modulation and Element-3

*Envelope modulation and Element-3*

As we said earlier in this training, Element-1, Element-2 and Element-3 are the audio parts of a subtractive synthesizer. The audio signal starts its journey at Element-1 and from there it goes into Element-2 and continues onto the final stage, Element-3, which is the amplifier. With synthesizers we are able to control a lot of parameters and in particular, the volume parameter is a very important one to have control over. Once we hit a key, we'll want to settle on the type of volume changes for our sound that we are creating. Are we going to have a slow rising pad sound or a short percussive dance sound?

Because of the fact that Element-3 is the last stage in the audio signal, Element-3 is also a final control over the sound. If we have any settings incorrect, a pad sound won't sound like a pad sound anymore.

The perfect tool to control a volume change in any synthesizer sound is an envelope. This envelope, assigned to the volume parameter of Element-3, is also the final control in how the volume will change of the sound. In the Element-3 chapter we've already explained the amplifier and also the volume envelope. The volume envelope is so essential, on most synthesizers we don't even see the real Element-3 parameters, but only the envelope that controls the volume. This reveals again how important the envelope is in conjunction with Element-3 and how it controls the volume parameter. So a volume envelope is a part that we'll surely find on all subtractive synthesizers.

Remember these points from the chapter on Element-3:

- The volume parameter is the initial volume; once we open it, it adds a continuous bleeding of the signal that comes from the filter Element-2.
- Envelope amount determines the volume of the sound; on all synthesizers this is called volume instead of envelope amount.

The term envelope amount inside the amplifier section is something we'll never see on a regular synth. It would be too confusing, it's labeled volume. So the volume control which we can see on the old Prophet-600 is actually the envelope amount.

Let's get back to the envelope which is an essential tool inside Element-3. We're going to repeat a bit of the part in which we explained the envelope.

**The attack stage**

- attack is the first parameter that starts to act once we hit a key.
- attack always goes from 0% level up to 100% level.
- if attack is set to 0 (zero) time, the envelope starts at 100% level from the moment we strike a key.

Figure 4.36

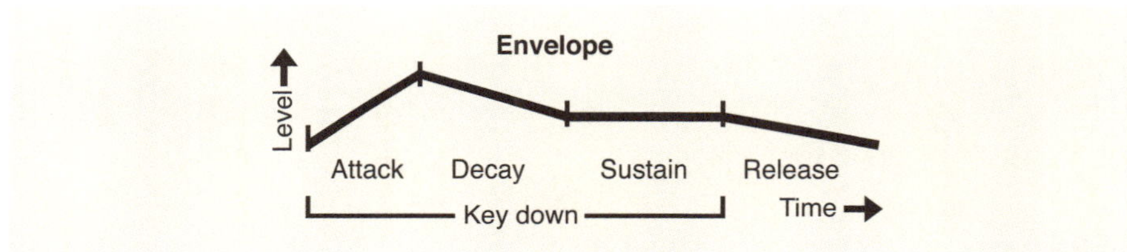


Attack from 0% level up to 100% level.

The envelope is connected to the volume and now this means that the level of the envelope controls the volume parameter. So using a slow attack time we can create a sound that rises slowly in volume. This is great for pads or other sounds that have no direct attack sound.

Figure 4.37

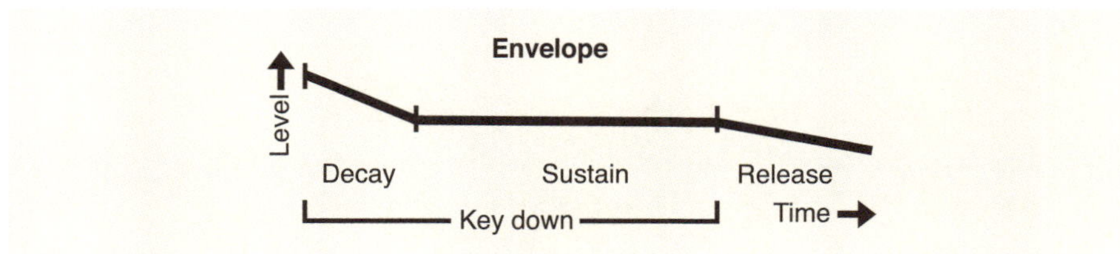


Attack starting at 100% level.

Having the attack time set to 0 (zero) is great for more percussive sounds like bass and many other sounds that don't use a slow attack. The envelope starts at 100% level from the very beginning. So the volume will be at max level from the moment we hit the key.

### The decay stage

- decay starts to act after the attack stage reaches 100% level
- decay is the first parameter that starts to work if the attack time is set to 0 (zero)
- decay always moves on to the sustain level
- decay does nothing if we have the sustain level set to 100%

### The sustain stage

- sustain starts to act after the decay stage ends
- sustain fixes the level of the envelope as long as we hold the key

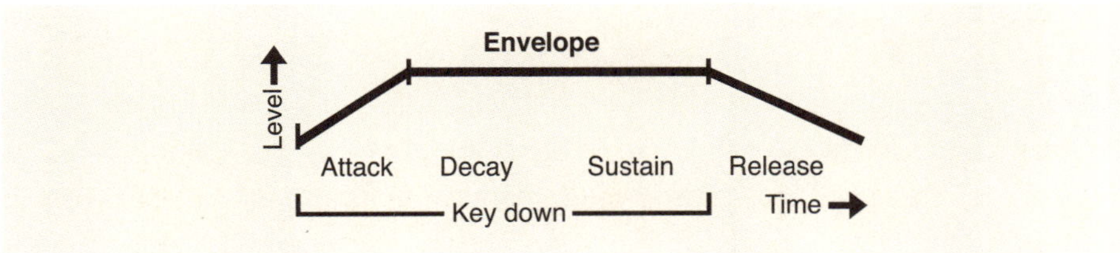


Figure 4.38

Sustain at 100% level.

With a pad sound we can hear very clearly that the sustain of the envelope holds the level of the volume. In fact, when we create a pad sound we have to keep the sustain at the 100% level, because if we set the sustain to a lower level the pad sound will collapse the moment we hold the key. If we keep the sustain set up to the 100% level, decay is not active because it has no level to fall to.

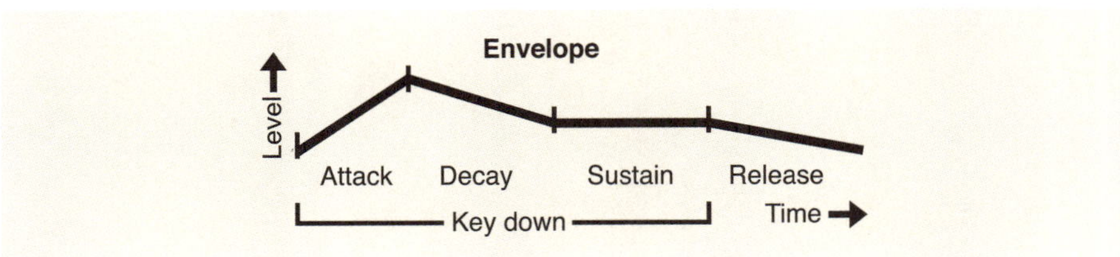


Figure 4.39

Sustain at 50% level.

But for instance with a brass sound, the sustain level can be lowered. This is because it will sound quite natural to have a lower sustain level on this particular type of sound.

Figure 4.40

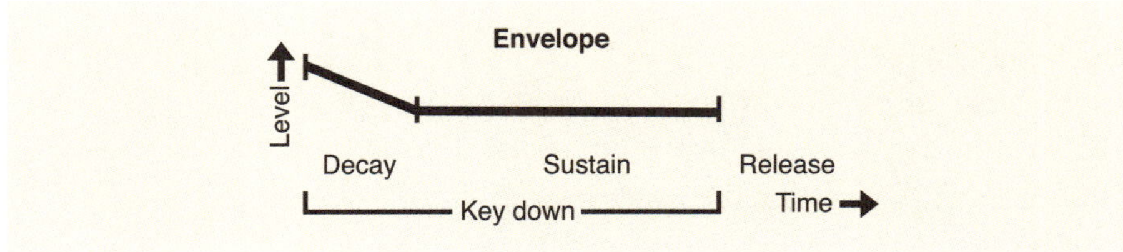


Sustain at 0% level.

With percussive sounds sustain can be left at 0 (zero): did you ever hear a snare drum that held its sound for one bar? Or a cowbell that can be sustained during the entire bridge of a song?

It is not a bad idea to look at the acoustic world and how things work there. We will find out that the sustain of the volume envelope is not really a viable option for percussion or drum sounds. In fact, even with plucked sounds and bass sounds it is the same story. Once we hit a string it's always going to decay, it never has a natural sustain.

As a general rule, the sustain can be set to 0% in the case of drums, percussion and many string type of sounds.

**The release stage**

- release only starts after we let go of the key
- release always descends to the 0% level
- if an envelope has already reached 0% level, release is not active.

Once we let go of the key, release starts to act, and the release stage will go to 0% level of the envelope. The envelope will control the volume, so in this case it brings volume down to 0% level. And in the case of a volume envelope this means silence. So after the release of the key, the envelope release stage is in charge. Only if we have sustain set to 0 (zero), and keep the note down as long as the decay stage is active, the release does not work.

***Keep in mind with release on a volume envelope that also the release on the filter envelope is active at the same time. So it's wise to keep the filter release a bit in line with the release of the volume envelope. Of course, if we use more envelopes inside our sound, we'll want keep up with what the other envelopes are doing.***

*Element-4: 'Envelope and Element-3'.*

*Envelope modulation and Element-4*

## Envelope modulation and Element-4

At this stage the 4ES chart is once again essential to understanding what we can do with a subtractive synthesizer if the features are available.

Element-4 shows several modules and the one we are discussing now is the envelope. The red dots on the chart are inputs. We can connect parts from Element-4 to other parts of the 4ES. But don't forget: Element-4 parts can be connected to other Element-4 parts! For example, we can easily connect the envelope to the LFO and let the envelope control its speed.

Figure 4.41

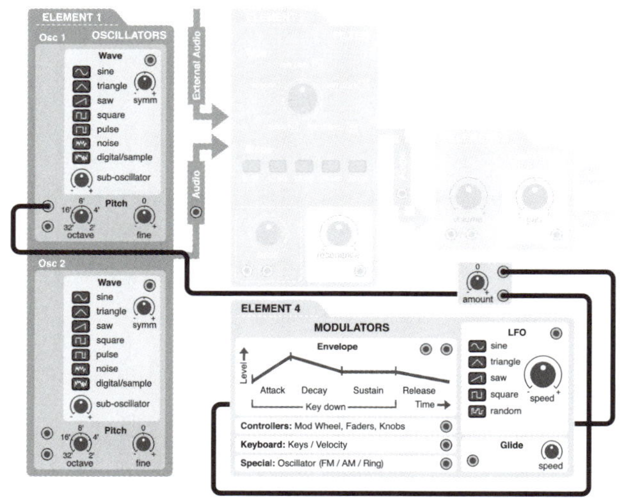


Envelope controls LFO amount.

One of the most common uses for an envelope is for it to control the amount knob. We can see that the amount knob also has a red dot which stands for input.

We use this for example when building a violin-type solo sound. What we would want is that the vibrato inside this sound is not really at full strength when the sound starts, it should fade in.

The way we accomplish this is by using an envelope to control the amount of vibrato produced by connecting an LFO to the pitch of an oscillator. So if we picture this using our 4ES chart, the LFO is routed to the amount knob and from there it's routed to the pitch of the oscillators. Then the envelope goes to the input of the amount knob, this will control the amount of signal coming from the LFO. So instead of having a fixed amount level, the envelope will control this level. In our violin-type solo sound the attack time is opened a bit and we'll have the vibrato increased while holding down the note.

Using an envelope to control the level of amount can be very useful and handy. Of course there are many more examples of things we can do by connecting the envelope to Element-4 parts. Another common connection is to use an envelope in combination with the vibrato speed to give the vibrato a more natural feeling.

WATCH ONLINE ***Element-4: 'Envelope and Element-4'.***

*Envelope trigger modes*

**Envelope trigger modes**

The very first synthesizers were monophonic. Monophonic means that only one note can be played at a time and the difference between some monophonic synthesizers was the way that the synthesizer responded to the keyboard, and especially legato notes. This is called trigger mode.

Trigger means that the envelope starts to work and goes through its envelope using attack, decay, sustain and release. The envelope trigger modes often worked differently on several brand of synthesizers. For instance, when we use the Korg MS-20 and play a note while holding down another one it does not retrigger the envelope. So in case we have a slow attack, we have to let go of the note before we can have another slow attack sound. This mode is called legato mode and it's used on several other monophonic synthesizers. What happens is that the envelope goes through its full envelope cycle, even if we hit another note while holding down the first note. The envelope only re-triggers if all the keys are released or if the release stage has gone back to 0% level. So if we have a long decay setting on our envelopes and the sustain is open to a certain level it allows the envelope to finish all its stages, even if we hit another note while playing legato.

With the entrance of polyphonic synthesizers the envelopes were triggered per voice. Each voice triggered its own envelope. A polyphonic synthesizer is in fact a stack of monophonic synthesizers that play the same sound. Each voice has its own full synthesizer with oscillators, filters, amps, envelopes, and so on. That's the reason why the first polyphonic synthesizers were so expensive. Inside a classic Jupiter-8 there are eight monophonic synthesizers, eight voices as they call it. So a Jupiter-8 has 8 x 2 oscillator = 16 oscillators, 8 filters, 8 LFO's, 16 envelopes.

Because of the popularity of the typical legato sound that we know from monophonic synth sounds, like MiniMoog basses and lead sounds, it is still available on today's synths. The legato mode is especially popular. What makes it so special on a vintage MiniMoog is that the filter envelope is also responding in a legato manner. So the filter closes using the decay parameter and adds this great sound of a changed filter even if we hit other notes.

**• Mono**

On other synths like for instance a Modular, we notice that every note is re-triggered. So legato mode is not active here. This mode is basically the mono mode, playing legato here makes the notes cut off. This mode means that the keyboard is triggering the envelope once we hit a new note, even while holding down another note. Every note triggers the envelope even if we're playing legato. This mode is great for sequence-type sounds.

**• Legato**

This mode means that the note is triggering the envelope once we hit the first note. Playing legato means that the envelope goes through its full cycle even if we hit another note before the cycle of the envelope has been completed. This mode is great for solo sounds and basses.

**• Poly**

This mode means that each voice inside a polyphonic synthesizer has its own envelopes.

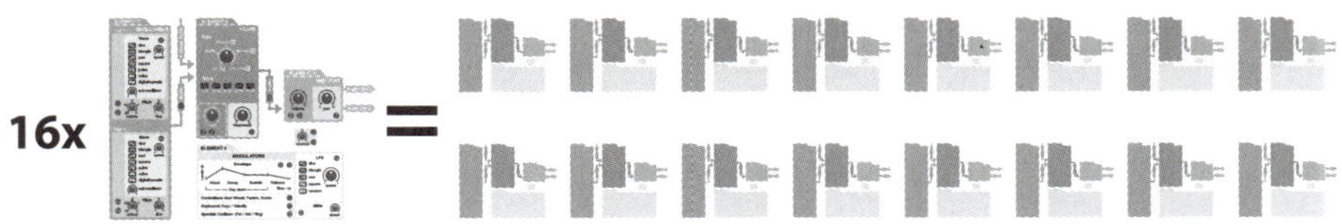


Figure 4.42

16 voice synthesizer.

Predator is 16 voice, which means that 16 synthesizers are active if we play 16 notes. All of the voices have their own oscillators, filters, envelopes, and so on. Legato playing means that we'll hear more voices at the same time.

*Element-4: 'Trigger/play modes'.*

*LFO*

**• LFO**

If you've been playing synthesizers for any length of time at all, we're sure you've seen the name LFO on the front panel. Of course, we'll find it on countless software synthesizers.

With lots of sounds, we can hear vibrato quite easily within the sound. Vibrato is a continuous pitch change that's utilized by singers, guitar players, it's used on countless acoustic instruments. To generate vibrato on a synthesizer we need something that will constantly continue to modulate. Unlike an envelope, the LFO is a module inside a synthesizer that has the ability to produce this effect.

***Once again, the essential key to Element-4 is:***
***it makes the other elements work and adds dynamic changes.***

This is exactly what an LFO does as an Element-4 part: the LFO adds repeating changes to the parameter we assign it to. The clearest example of this fact is having the LFO connected to the oscillators for generating a classic vibrato effect.

Figure 4.43

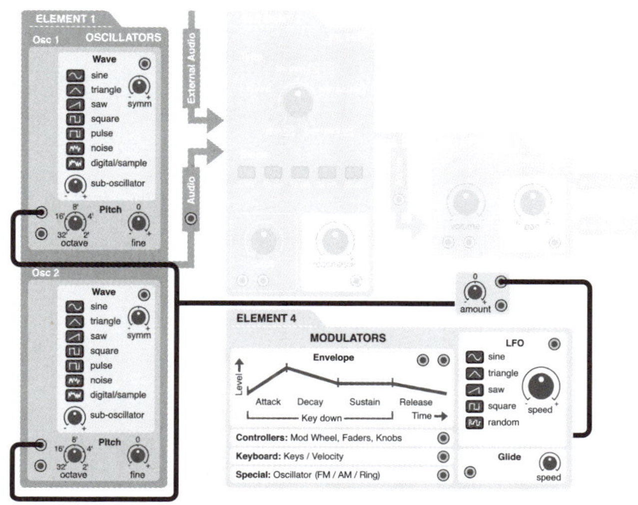


LFO connected to the oscillators.

The Mini-moog has the option to set the oscillator at 'low' frequency. LFO stands for Low Frequency Oscillator, and the symbols of the waveforms we've seen with regular oscillators now display the type of modulations available to us. We can switch from sine to square and we'll notice it makes different types of modulations. Of course for vibrato we have an amount knob in between the LFO and oscillator, and for vibrato we use the sinus waveform or triangle waveform. We could use one of the others, but it's not going to sound as useful for vibrato.

There's another typical thing that happens once we use an oscillator as an LFO. The frequency setting that we would normally use to fine-tune an oscillator is still the same parameter, but once we change an oscillator to a Low Frequency Oscillator, the frequency parameter now becomes the modulation speed of the LFO.

If we look at the Access Virus, the speed knob in the LFO section is actually called rate. Looking at the Prophet-600, it's still called frequency. Looking at Predator, LFO uses 'speed' instead of 'frequency'. Speed is easier to understand than frequency, and the term also makes sense because we can't turn these LFO's on most synths into normal oscillators.

Let's recap:

- LFO stands for Low Frequency Oscillator.
- LFO is a normal oscillator, but one that plays at a very low frequency.
- LFO waveforms use the same oscillator waveforms, but instead of producing a sound, they produce different types of modulation movement.
- LFO on many panels say speed but they are actually referring to frequency.

Most synths actually use LFO as the name for their LFO's, but sometimes we still may need to grab our manual because, for example, if we look on the classic Korg MS-20 there's no marking for LFO. Instead of LFO we have to search for a part that looks like an LFO. It's called MG which means Modulation Generator as we can find out in the manual. Not a bad name of course, but this MG is just an LFO.

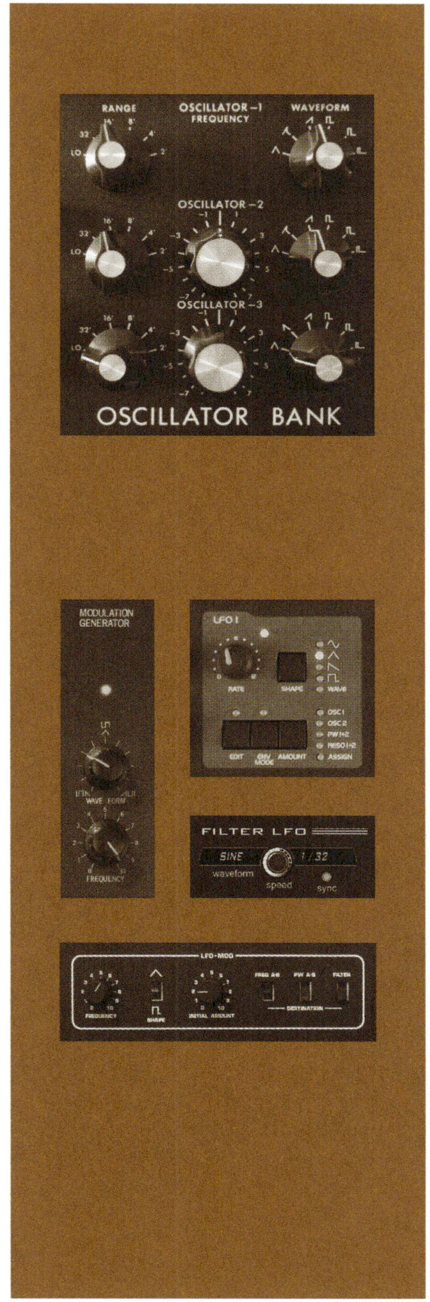

If we look at a Mini-Moog there's also no LFO. The Mini-Moog has three oscillators and the third one is mostly used as an LFO. The Mini-Moog was one of the very first stage performance synthesizers. Before that only big systems were available.

So Moog built this first small synthesizer with stage use in mind without a dedicated LFO. Later on, most builders used dedicated LFO's inside their synths because they were needed all the time. It became standard practice to have an oscillator dedicated as a Low Frequency Oscillator and to rename frequency to speed. The LFO is an interesting Element-4 part, and as we said before, its most obvious use is for pitch vibrato. But if we take a look at the 4ES chart we can see there are many inputs that we can connect the LFO to.

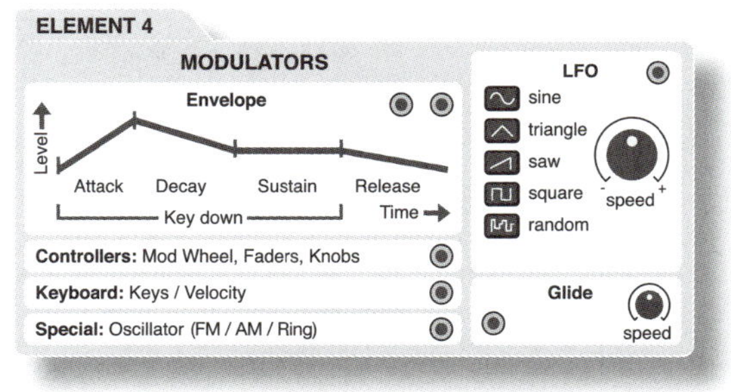


Figure 4.44

*Element-4: 'LFO / Introduction'.*

**• LFO speed**

In the introduction on the LFO we stated that the speed parameter of an LFO is basically the frequency of the oscillator. So once we change the frequency we can hear that the speed of the modulation also changes. Most synthesizers use the name speed or rate because that name best describes what's really going on.

*Element-4: 'LFO / LFO speed'.*

**• LFO speed modulation**

With the speed parameter we set the modulation speed and most of the time this is a static setting. If we look at the 4ES chart we see that LFO also has an input. Indeed this is a connection to the speed of the LFO, or the frequency of the LFO.

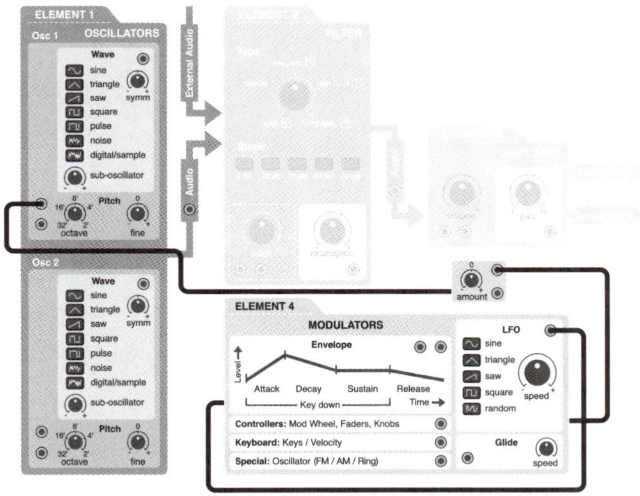


Figure 4.45

LFO connected to an envelope.

For instance, we can connect an envelope to it, this way we can generate a more natural speed change. With this connection we can control the speed of the LFO instead of running at the same speed all the time. This will allow us to create more natural and alive lead sounds. There are several more options we have in controlling the speed by using other parts of Element-4. Using today's synthesizers we can lock the speed of an LFO to the music. Instead of setting the speed in normal values, it can be set to musical values. By the bar, half a bar, 1/8 note, and so on. This is a great feature, and gives us extra sonic dimensions using the LFO!

*Element-4: 'LFO / LFO speed modulation'.*

**LFO waveforms**

Because of the fact that an LFO is an oscillator, we can find all the various waveforms inside of it just like any other oscillator. Each has its own character and some are commonly used, others not so much. The most commonly used waveforms are the sinus and triangle. Both are very similar to each other, and both work great for vibrato. The sinus and triangle waveforms are also great for PWM modulation. The other waveforms are more useful for FX sounds. The symbols are very close to what we actually hear, they describe very well the contours of the modulation they produce.

***Element-4: 'LFO / LFO waveforms'.***

**• LFO waveform: sinus**

Overall, the sinus is a nice and smooth up- and down-movement waveform. Its symbol displays this characteristic very well. It's very popular and it's one of the most heavily used waveforms for LFO along with the triangle waveform. The sinus is great for vibrato if we connect the LFO to the pitch of the oscillators. It's also great for waveform modulation of the oscillators which is often called PWM modulation. Another use is filter cutoff frequency modulation. Finally, it can also be used for panning modulation or for a tremolo effect on the amplifier.

***Element-4: 'LFO / Sinus, saw and pulse'.***

**• LFO waveform: triangle**

Overall, the triangle is also a nice and smooth up- and down-movement waveform. The symbol displays this characteristic very well. It's one of the most heavily used waveforms just like the sinus waveform. The differences between the sinus and triangle are hard to hear, so basically we'll get the same results as the sinus if we use the triangle as an LFO waveform. The triangle is great for vibrato if we connect the LFO to the pitch of the oscillators. It's also great for waveform modulation of the oscillators which is often called PWM modulation. Another use is filter cutoff frequency modulation. Finally, just as with a sinus, it can also be used for panning modulation or for a tremolo effect on the amplifier.

***Element-4: 'LFO / Sinus, saw and pulse'.***

**• LFO waveform: saw**

We have the Saw waveform in two versions, saw up and saw down. The saw up is also often called ramp. If we select these waveforms inside a normal oscillator there's no difference between them in sound, only the polarity of the waveform is reversed. This reversal can only be heard when we use the oscillator as an LFO.

Figure 4.46

It's basically only useful for FX sounds, when it's assigned to the filter or amplifier it sounds like a repeating envelope. Like we mentioned in the LFO speed section, it is possible to synchronize an LFO to the tempo of the music, so then the speed parameter has a musical value. With the saw waveform as a modulator, it will generate rhythmic sound effects.

***Element-4: 'LFO / Sinus, saw and pulse'.***

• **LFO waveform: square / pulse**

The sound of the LFO square waveform or LFO pulse waveform is just as its symbol reflects. It is like an up and down switch in sound. A regular LFO found on most synths will offer the square waveform. Remember from an earlier chapter that we said a 50% pulse is the same as a square waveform.

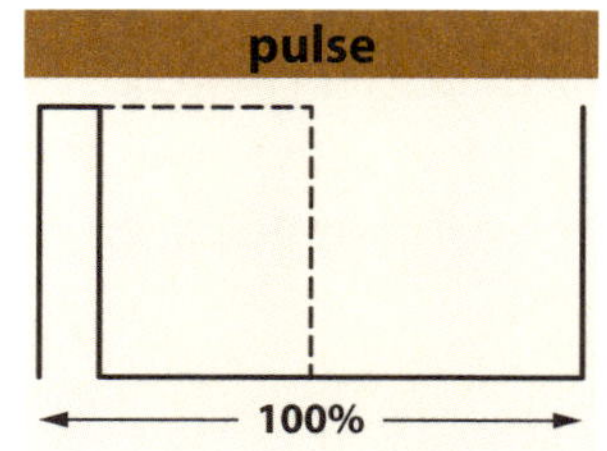


Figure 4.47

When using the Pulse waveform we can change the width of the pulse. We could, for example, set the width to 10% which would result in a narrower sound. Well, if pulse width control is possible inside an LFO of a synth, then the pulse waveform will give us a different sound when it's used as a modulator if we use a value other than the regular 50% pulse width. We can do this on most modular synths because often there we are using a regular oscillator as our LFO. The LFO square waveform is especially useful for sound effects.

***Element-4: 'LFO / Sinus, saw and pulse'.***

• **LFO waveform: random or sample & hold**

The random waveform of an LFO is the only waveform that is different from a regular oscillator. It is in fact pink noise but by using an additional connection of the LFO it turns into sample & hold or random. Let's explain this in greater detail using the typical setup of a sample & hold generator. First, we take pink noise and feed it into the sample & hold section. Then, we pick the pulse waveform output of an LFO and connect it to the clock input of the sample & hold.

Finally, we connect the output of the sample & hold section to the modulation input of an oscillator. This way the pink noise is not fed directly into the modulation target. The pink noise is generating a random uncontrollable modulation. What the LFO pulse does is chop the pink noise into small values. The best waveform for this chopping job is the pulse waveform because it produces an on-and-off switch effect. The symbol of the square reflects this on/off effect quite well.

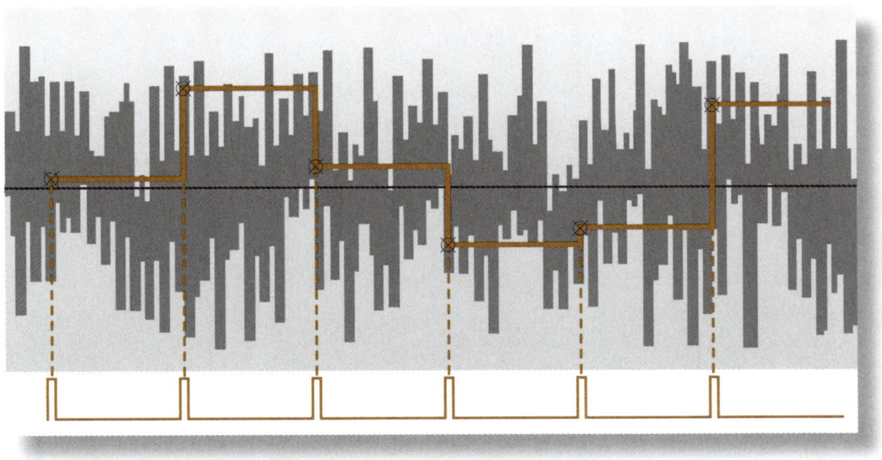

Figure 4.48

We can easily follow the modulation created by a sinus waveform, but regular pink noise has so many fast jumps in its output that we can't hear the individual jumps. So we use the pulse waveform to chop up the signal and sample short values of it. Therefore, we need to set the pulse waveform here to narrow. Otherwise we'll get too many changes or noise in the modulation signal. It really needs to just sample a very short fragment of the pink noise. This short sample is then a random level and when it's connected it produces these typical random effects. Now we know how 'sample & hold' or random, as most synth brands call it, is made. But what can we do with it? Next to being great for sound effects, it is also interesting as a waveform if the LFO can be synchronized to the tempo of our music software.

***Element-4: 'LFO / Random and sample & hold'.***

## LFO waveform modes

Many of today's synthesizers also have waveform modes or LFO modes as options for their LFO's. What do these modes mean? Let's look into this. Remember that the LFO is just a regular oscillator that plays at a low frequency and can be used as a modulator. If a synthesizer has enough voices to play 16 notes at the same time, then this means that for each note we play a complete monophonic synthesizer is playing. So then a 16-voice synthesizer can be seen as 16 monophonic synthesizers playing the same sound.

When using an LFO we can decide to use one LFO for all these voices or one LFO for each voice. This is what we're talking about when we say LFO modes. Older analog synthesizers and also many digital subtractive synthesizers just use one LFO for all the voices of the synthesizer. This is the most common use of the LFO. In this digital age we have the ability to decide to use one LFO for all the voices, or to give each voice its own LFO. If the option is there to have an LFO for each voice, it can be cool, but it's not always needed.

### • Free mode or mono mode

The first mode is the free mode, which is the classic mode. This mode is used on most all classic synthesizers. One Low Frequency Oscillator is used to generate the modulation. The oscillator runs free, which means that the waveform does not reset when we hit a note. And, even if we play more notes, they all get the same modulation from this LFO. This mode is great for vibrato and can be found on all classic analog synthesizers. Also, when we say "one LFO" we mean mono mode.

Figure 4.49

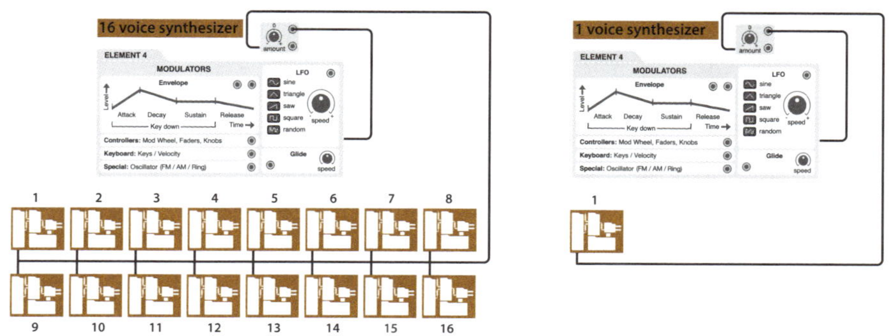

**• Mono mode with reset**

This mode is also known as reset mode because the LFO's waveform is reset each time we hit a new note. The waveform starts back at zero, or the beginning of the waveform. Once we play, we can hear it very clearly. Each time we hit a new note or add another note, the LFO waveform resets. This mode is great if we want to have more control over the movement of the waveform. Especially if the LFO can be midi tempo based, interesting results are possible. The LFO will retrigger each time a note is hit.

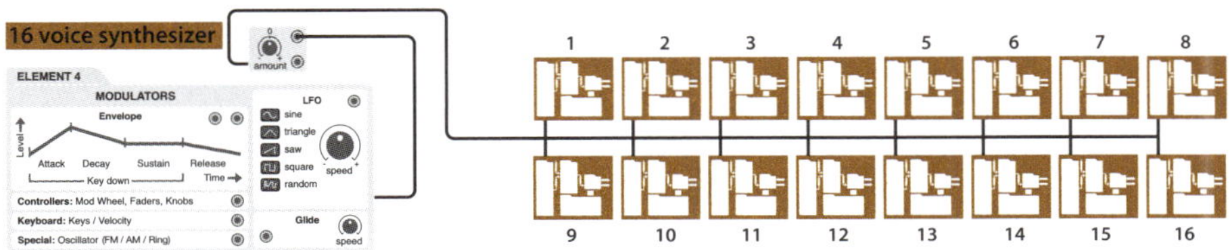

Figure 4.50

**• Poly mode**

This mode is called poly mode because each voice of the synthesizer gets its own LFO. This is set up so that once we play more than one voice on our synth, we can hear the individual LFO's playing. This poly mode is great if we want a "swirly" sound. For instance, we use two oscillators and on the second oscillator, we add a bit of LFO modulation. The sound becomes fuller by using poly mode because each voice has its own independent LFO running and starting on its own time.

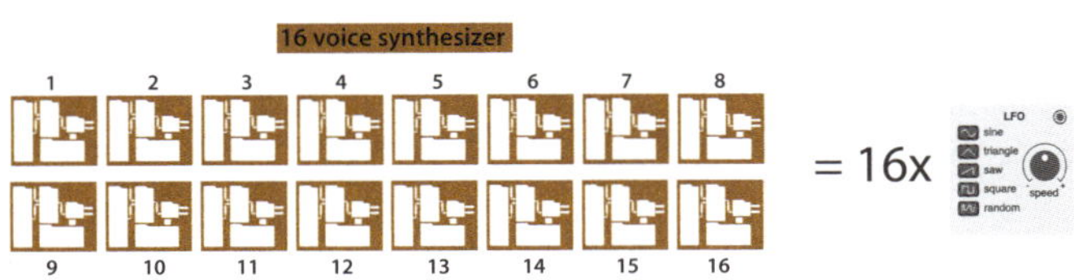

Figure 4.51

These modes are very interesting sound making tools, especially if we use them in combination with midi-tempo based LFO options.

WATCH ONLINE *Element-4: 'LFO / Waveform modes'.*

*LFO and Element -1*

## LFO and Element -1

If we look closely at the 4ES chart, we see that an oscillator consists of two components. These two components are called pitch and wave. The pitch component of the 4ES oscillator has two inputs. This is symbolic to show that this is commonly used. As we're sure you will recall, we connected the keyboard to this pitch component of the oscillator so that we'd be able to play melodies.

On the wave component of the oscillator we see only one input. This means that the modulation of the wave component is less often used compared to the pitch component.

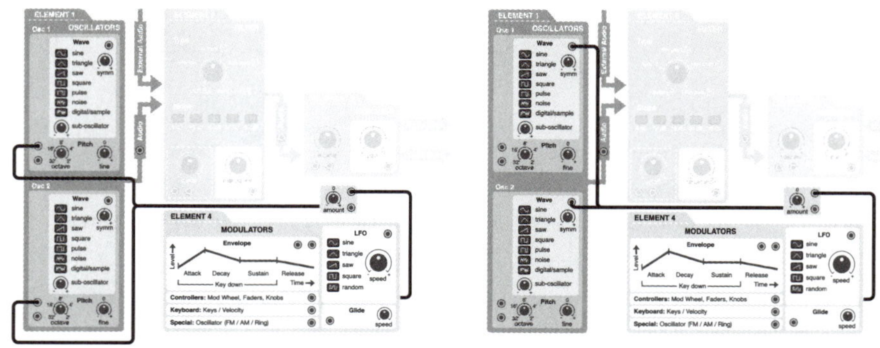


Figure 4.52

We've already explained the LFO a bit and this Element-4 part can be connected to the pitch and wave component of Element-1. Vibrato is a very clear example of what we can do with an LFO controlling Element-1. We create vibrato when we connect an LFO to the pitch component of an oscillator.

## LFO and pitch modulation vibrato

Let's review the LFO to pitch modulation connection by using the 4ES synth. The LFO goes to an amount knob, which allows us to have control over the amount of modulation. From there the LFO goes to both oscillators and their pitch input. This is how we create vibrato.

The sinus and triangle waveforms are the most commonly used for vibrato. Of course, we can use the other waveforms for vibrato, but then we'll hear that it's not a natural vibrato anymore but more like a sound effect. The pulse waveform is rather special in that it adds a kind of trill to the sound that can be cool for special leads.

### LFO and pitch modulation vibrato controlled

On for instance the Mini-Moog, the connection between the LFO and the oscillators for generating vibrato is already set. The modulation wheel on the Mini-Moog is used as the standard amount control for this. On many other synthesizers this connection is different. Often we have a regular amount control and we have to set up a different connection to use the modulation wheel or even aftertouch as the amount control. The regular amount control in between the LFO and the oscillators allows us to always have a certain amount of vibrato. The vibrato is playing and is heard all the time. To be able to have more expression though, an extra amount control needs to be added. This means that we want to use another Element-4 part to control the amount. In this case, we're using the modulation wheel so the amount is now controlled by the modulation wheel as well.

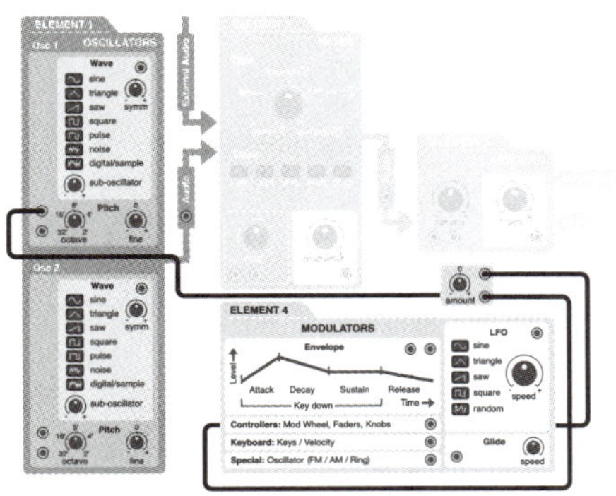


Figure 4.53

Most of today's synthesizers have the connection as described on the previous page. This way, there is always a basic amount of modulation and also an external control possible. We are, of course, free to connect even more Element-4 parts to the pitch of an oscillator or even to the LFO itself.

*Element-4: 'LFO and Element-1 / Introduction'.*

*Element-4: 'LFO and Element-1 / Pitch modulation FX sound'.*

**LFO and pitch modulation tip**

With vibrato we usually modulate both of the oscillators. But for instance on a Jupiter-8 we can select only one of the oscillators to modulate. In the case that both oscillators are being used for a sound and we want to have vibrato, then it's going to be necessary to modulate both of them.

But for making sounds that will be a bit more alive, it is also nice to modulate only one of the oscillators slightly instead of both. This thickens the sound up. If we modulate it too deeply it won't sound good. So, by tweaking the amount we can get just the right setting. This can be compared to a chorus effect. Only instead of using a chorus, we use the oscillators and an LFO.

*Element-4: 'LFO and Element-1 / Pitch modulation tip 1'.*

*Element-4: 'LFO and Element-1 / Pitch modulation tip 2'.*

### LFO and waveform modulation

Let's take another look at the 4ES chart and Element-1. The second component of an oscillator is the waveform. Here we can select the color of the sound of an oscillator. And here as well, we can see that there's an input for us to connect parts from Element-4 into this waveform section. In the earlier chapters about Element-1 and in particular, the pulse/square waveform, we discussed this option to modulate the waveform. In fact, what we're really doing here is modulating the symmetry of a waveform.

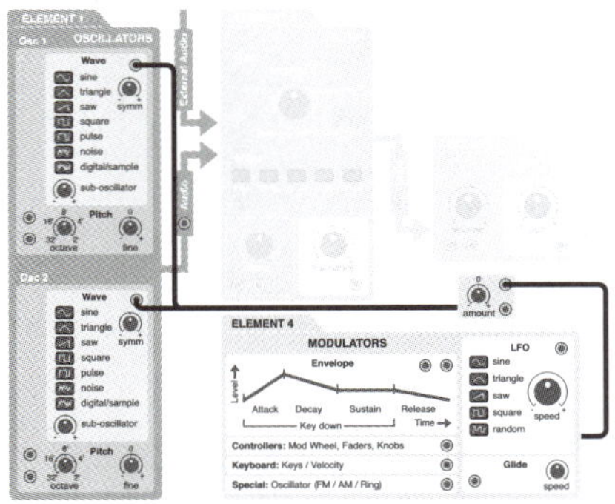


Figure 4.54

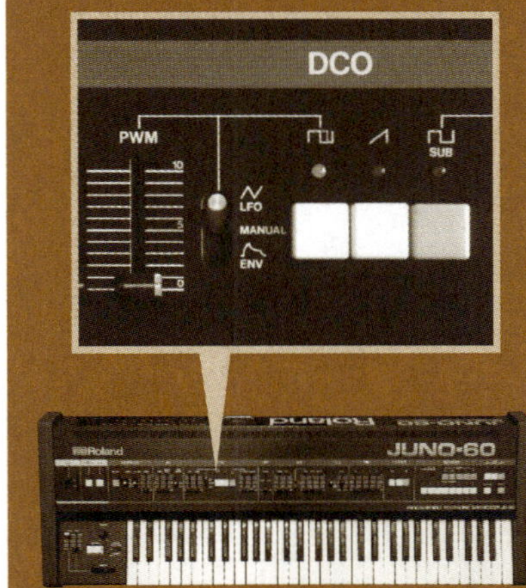


The modulation of the LFO on the pulse width is a great tool. It produces a sound that is moving and "chorusy" together. That's a big difference from the more static sound of the saw waveform. This chorus like sound is great for building pad sounds or heavy hard-style dance-lead sounds. The difference between these two sounds is how deeply the LFO is modulating and at what speed. So, deeply applied and high speed LFO modulation is great for sounds like a hard style dance-lead and more subtle modulation is great for pad sounds.

WATCH ONLINE ***Element-4: 'LFO and Element-1 / Waveform modulation'.***

*LFO and Element -2*

## LFO and Element -2

In the chapter about Element-2, the filter, we learned that cutoff frequency and resonance are the parameters that we can modulate.

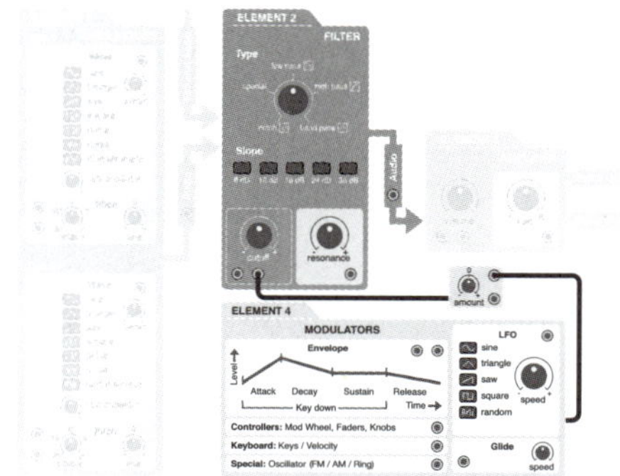


Figure 4.55

When using an LFO to modulate the cutoff frequency, instead of a single movement like an envelope , we create endless modulation. With Element-1 the LFO is often used for natural vibrato. But with Element-2 there is no such classic icon in how the sound changes. Vibrato, we can easily recognize from being used on acoustic instruments or being performed by a singer. It's much different when we are talking about the LFO connected to the filter. The closest examples would be a repetitive wahwah effect on a guitar or the sound that a trumpet player produces if he uses a mute. Still, using the LFO for filter modulation is commonly used. Especially if the LFO can be midi tempo based. Of course, we have to keep in mind that the amount setting and the cutoff frequency are very important. Remember that the cutoff frequency is the initial setting at which the modulation is going to happen.

### LFO and resonance modulation

If we look at the 4ES chart we can see that we also have the option to modulate the resonance. What we actually modulate is the amount of resonance.

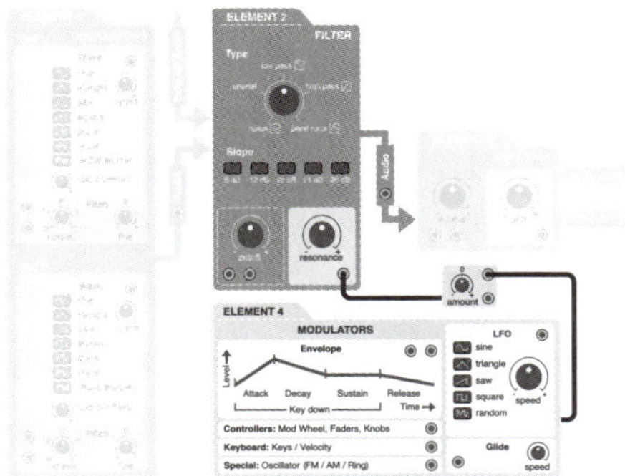


Figure 4.56

Overall it is rarely used, so doing this with an LFO is also rare. But, it can still generate some very interesting sounds.

***Element-4: 'LFO and Element-2'.***

## LFO and Element-3

Element-3 has one major parameter, which is volume and on many of today's synthesizers we can also find a panning parameter. Of course we can modulate both using an LFO. The volume modulation is known in the acoustic world as tremolo, which you may have heard as an effect used by guitar players. It is a continuous volume movement.

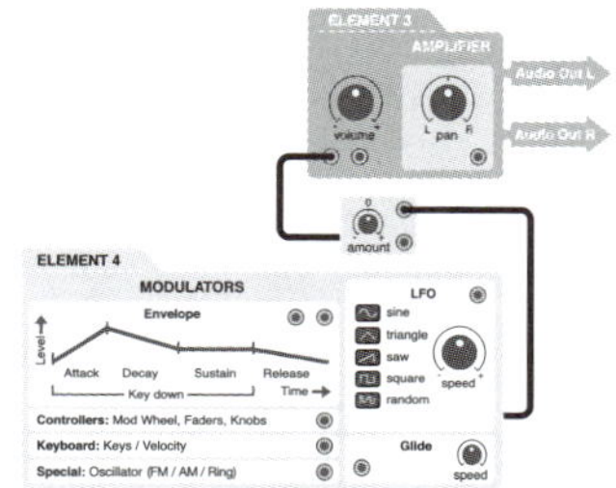


Figure 4.57

The LFO modulating the volume parameter is rarely used. In the case of musical usages for this, think about tremolo for e-piano sounds. Another option is to use it for creating effect sounds. Today's modern synthesizers give us the ability to synchronize the LFO at the tempo of a sequencer. This adds interesting possibilities for modulating the volume using an LFO.

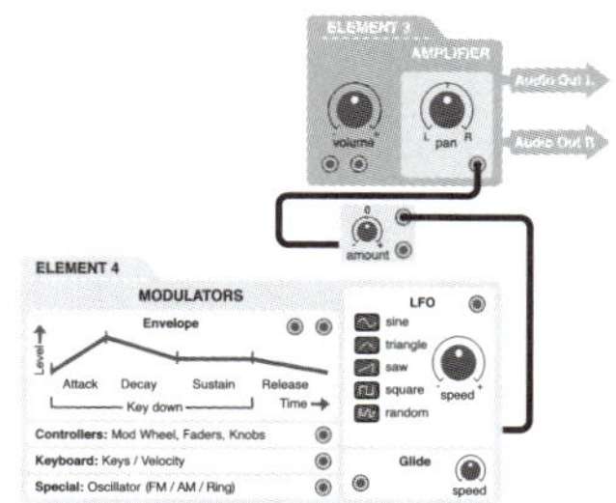


Figure 4.58

Modern synthesizers offer this option as well as the option to modulate the panning of the amplifier. This can generate interesting moving sounds that range from effect sounds to musical sounds that move around the stereo image.

WATCH ONLINE ***Element-4: 'LFO and Element-3'.***

## LFO and Element-4

*LFO and Element-4*

The 4ES chart tells us that parts within Element-4 can also modulate themselves. An LFO can be used to control another LFO! Or an LFO can be used to control the times of an envelope. Most of these are rarely used, but knowing the option may trigger some inspiration to use LFO's more in a different way.

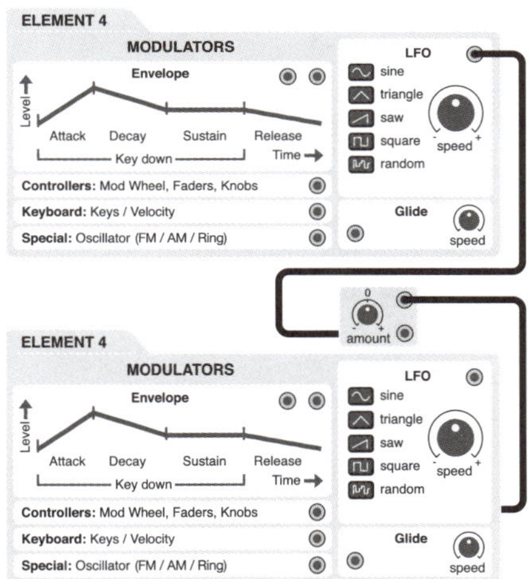


Figure 4.59

*Element-4: 'LFO and Element-4'.*

Keyboard

## Keyboard

The next part of Element-4 we're going to explain is the keyboard as a controller. Notice that as we are playing the keyboard we're selecting what notes to play, as well as deciding how hard we're going to play the keys.

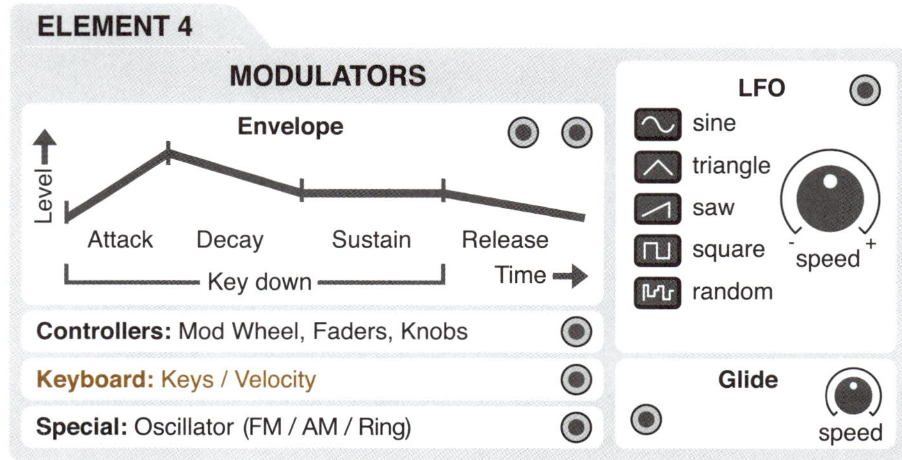


Figure 4.60

This is quite different from what we saw the LFO and envelope doing, which are also Element-4 parts. The human factor is important with LFO's and envelopes, because we make the decision on where to set the parameters like attack, decay, LFO speed, and so on. But that is indirect control. With the keyboard as Element-4 part the action is real time human input.

Once again the essential key to Element-4: it makes the other elements work and adds dynamic changes.

So in the case of the keyboard the dynamic changes are caused by us directly. As the player we're going to decide how hard we play the keyboard and which note or notes we're going to play. With keyboard as Element-4 part, we humans are generating these dynamic changes directly. And the fact remains that if we don't play, the other element-4 parts won't begin to act as well.

What if a sequencer or arpeggiator is playing? These parts take over partial control of the keyboard function and are a kind of replacement for the keyboard. What does the keyboard offer us? The most common and essential thing is of course the pitch and trigger information that it provides, so that we can hear the notes we play. Next to that it provides how hard or soft we play the keys, this is what we call the velocity information of the keyboard. Some midi keyboards offer release velocity, which means that they provide the information on how fast we release a key. These particular keyboards not only provide the information about how hard our fingers land on the keys, but also how fast or slow we release these keys. To use that feature will depend on whether or not our synthesizer offers it and we'll talk about that later.

The essential functions of the keyboard are:

- the keyboard provides the note information
- the keyboard provides the velocity information
- the keyboard provides the release velocity information (if available)

There is yet another type of keyboard information which is available on many keyboards. It's called Aftertouch. This data is generated when we press harder on the keys while sustaining them. It's great for controlling vibrato. Aftertouch is related to the other controllers and can be placed in the category next to pitch bend and modulation wheel.

There are two types of aftertouch that a keyboard can produce and the first one is by far the most common: monophonic aftertouch, which we usually just call aftertouch. Monophonic aftertouch can be activated by pressing on only one key. We can hold down many keys and all the active voices will get the same aftertouch applied to them. So it allows the voices and notes to get the same effect, even if we activate aftertouch only on one single note.

The second type of aftertouch is polyphonic aftertouch which is very rare. Only a limited amount of keyboards provide this information. Each note has its own aftertouch, so in case we play a chord and use aftertouch only on a certain note, then only that note will have the aftertouch applied to it. A synthesizer needs to be able to respond to poly aftertouch, so we're going to need not only a keyboard that produces this polyphonic aftertouch but also a synthesizer that can respond to it. Most modern software synthesizers and many hardware synthesizers will respond to it.

Here's the complete list of functions we find on almost all keyboards:

- the keyboard provides note information
- the keyboard provides velocity information
- the keyboard provides aftertouch information

The keyboard functions that rarely appear:

- the keyboard provides release velocity information
- the keyboard provides polyphonic aftertouch information

In the next paragraphs we're going to show in greater detail what we can do with the keyboard. You may find out about some things that you didn't think were possible using the keyboard.

***Element-4: 'Keyboard introduction'.***

***Element-4: 'Sequencers and arpeggiators'.***

**Keyboard note information and Element-1**

*Keyboard note information and Element-1*

We've used this essential connection from the Element-1 to the keyboard to explain a very basic rule of subtractive synthesis. A subtractive synthesizer is built up using separate elements and then they are connected to each other.

The connection of the keyboard is needed to provide the oscillators with note information from the pitch component. Otherwise we would be unable to play music. Looking at the 4ES chart we see that we can connect the keyboard to the pitch as mentioned and also to the wave component of an oscillator. The connection between the keyboard and the pitch component is of course the most obvious and essential. We'll dig deeper into this subject in the upcoming sections.

**Keyboard note information at oscillator pitch (the basic connection)**

As we can see on the 4ES chart, the connection from the keyboard is going directly to the pitch input of an oscillator. If we were to put an amount control in between this connection, like we did with LFO's and envelopes, we'll end up having a tuning problem.

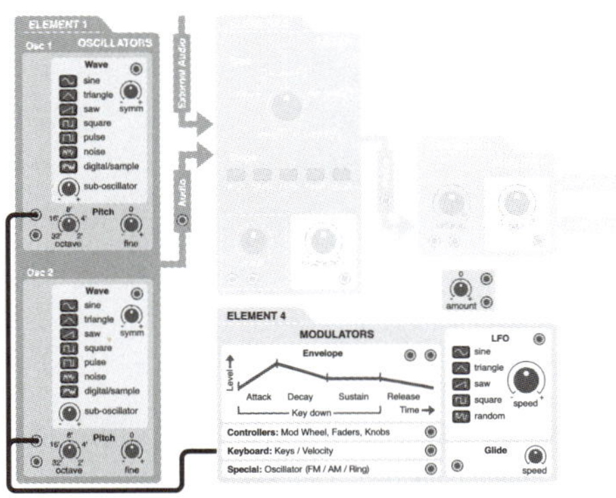


Figure 4.61

Suppose instead of connecting directly to the pitch input, we'll place an amount knob in between. In that case the notes are not in tune. Actually, they would sound very much out of tune. The amount needs to be set to 100% to keep the notes in tune.

So, for this reason the keyboard note connection is generally a direct connection to the pitch input and it's unlikely to have an amount knob in between this connection such as we have when using an LFO or envelope. Still, for some effect sounds, an amount control can be useful and even to have no keyboard connection at all. Let's look for this connection between the keyboard to the pitch component of the oscillators on some synthesizers. The connection between the keyboard to the pitch of the Predator oscillator is enabled by a button called TRK or track. This stands for keyboard tracking. Another description that we often see used for this function is called key follow. Both the names reveal that this switch turns the pitch tracking or pitch following on or off for the keyboard. So on Predator it is a button that connects the keyboard or disconnects it.

In general, we usually see only on/off connections on synthesizers to connect the keyboard note information to the pitch of an oscillator. Why should we set the connection of the keyboard to off? One example is when we want to create a percussion sound. For example, if we want to hit any key on the keyboard and get the same sound we'll set the connection to off. What will happen then, is that the initial setting of the pitch sets the pitch and it will remain the same over the entire keyboard.

For instance when we make a kick sound we can set the keyboard connection to off. This way we are sure that the sound is played at the right pitch, regardless of what note we play on the keyboard. We control the initial pitch with this semi-tone knob and fine-tune knob of the oscillator. Using these parameters we can find just the right pitch we want for this electronic kick.

On the other hand, if we turn on keyboard tracking in this case, we can use the keyboard to find a nice pitch for this kick, since now the pitch will change depending on what note we play. Also for FX sounds it can be handy to turn the keyboard note connection off.

***Element-4: 'Keyboard note information / The basic connection'.***

### Keyboard note information at oscillator wave

Looking at the 4ES chart reveals us that next to pitch we can also connect parts into the wave modulation input of an oscillator. Modulating the waveform means that it changes the symmetry of a waveform. We've already covered this in the chapter about Element-1.

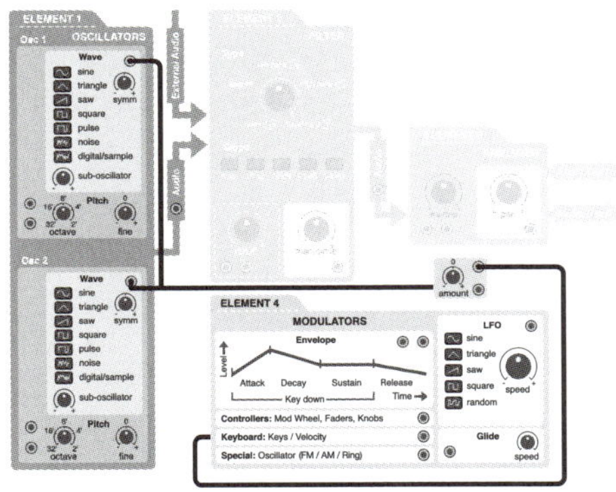

Figure 4.62

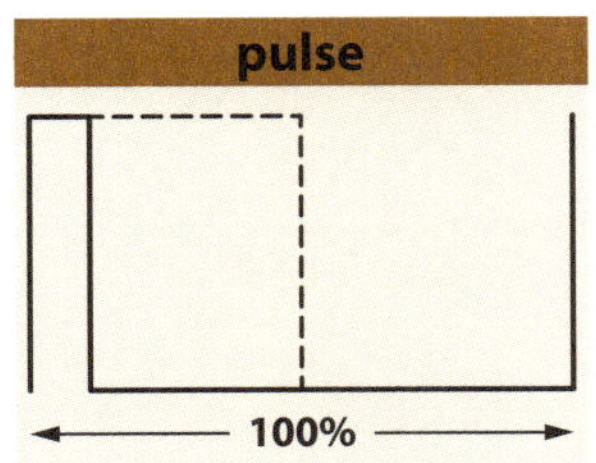


Figure 4.63

The most common modulation is changing the symmetry of the pulse waveform. This is also known as pulse width. The other classic waveforms like the saw waveform can be modulated as well but won't give us the same useful results like the pulse waveform will. We can modulate a waveform using the keyboard note information as a modulation source as well. For instance, inside the modulation matrix of Predator we can connect the keyboard to the symmetry of oscillator one's waveform. Inside the modulated oscillator we select the square waveform.

Once we open the amount, the keyboard note information changes the symmetry of this square waveform. Actually, it changes the pulse width of this square waveform. The higher we play, the more nasal the sound will become or in other words, the narrower the pulse waveform becomes. We can also choose another waveform and listen to how it's affected. We're going to hear changes in the sound, although the outcome will be different each time because of the color of the waveform.

This keyboard note connection to the waveform is rarely used, but sometimes the results can be interesting, especially when using digital or spectrum waveforms. It's good to experiment with the amount setting as well.

WATCH ONLINE ***Element-4: 'Keyboard note information / Wave modulation'.***

## Keyboard note information and Element-2

*Keyboard note information and Element-2*

The connection of the keyboard to Element-2 is often overlooked although it's used on nearly all sounds. If we take a closer look at several synthesizers and their filter panels, we can find the parameter named 'key follow' or 'key tracking'.

For instance on the Virus it says 'key follow', or on the Prophet-600 it says 'keyboard' and uses a 3 position switch inside the filter panel section, or on the Jupiter-8 it's called 'key follow'. The Mini Moog has no knob for it but instead uses switches for adding keyboard control to the filter. We can conclude from looking at all these synths that the keyboard is an essential modulator of Element-2.

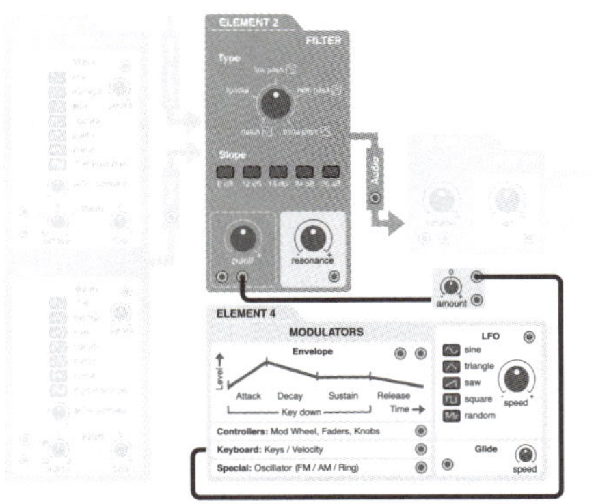


Figure 4.64

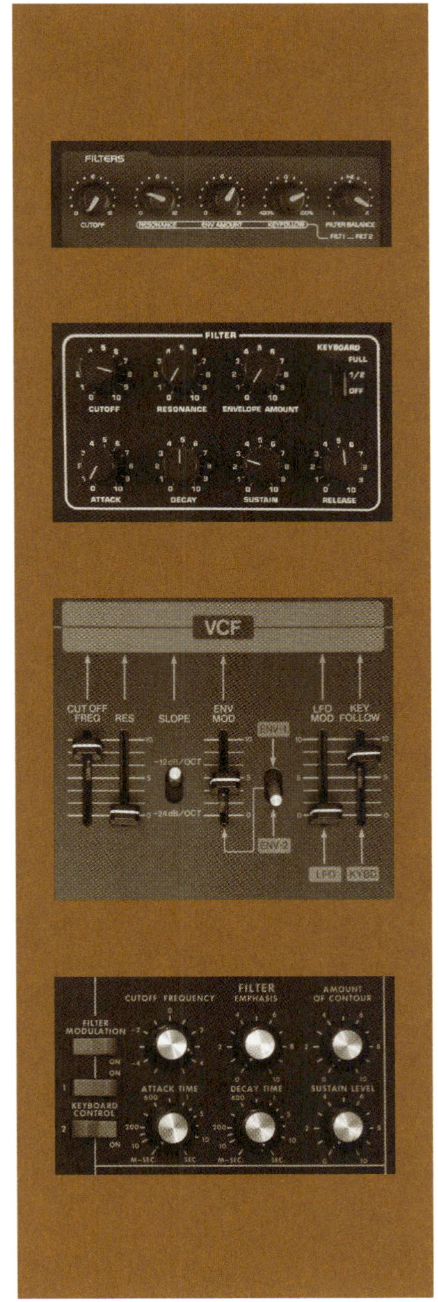


But which parameter is modulated by the keyboard? If we look at the 4ES chart, we'll see two parameters which have inputs that we can use for modulation, the cutoff frequency and the resonance. If nothing else is mentioned inside a synthesizer user guide, we can safely assume that the cutoff frequency is modulated by the keyboard note information.

We've already discussed resonance in the chapter about Element-2, and it's seldom modulated.

Now we know that it is the cutoff frequency parameter which is modulated by the keyboard note information. We also know that this modulation appears on almost all synths. So let's now explain why it is needed and also why it's such an important modulation parameter of the filter cutoff frequency.

First let's look at the acoustic world and how things work there. This is a good idea to do because many of the rules we can find there apply also to the synthesizers we play. If we play an instrument, say a piano in the lower keyboard range, the amount of higher harmonics are softer than if we play in the upper keyboard range. In other words, the higher we play on an acoustic instrument the higher the volumes become for the higher harmonics.

On our synth, the cutoff frequency in a regular low-pass filter setting takes off the higher harmonics. But we need to do this depending on the pitch range we play at. Remember that the cutoff frequency and its knob or parameter is the initial setting. So there's no information which key we're going to play!

Figure 4.65

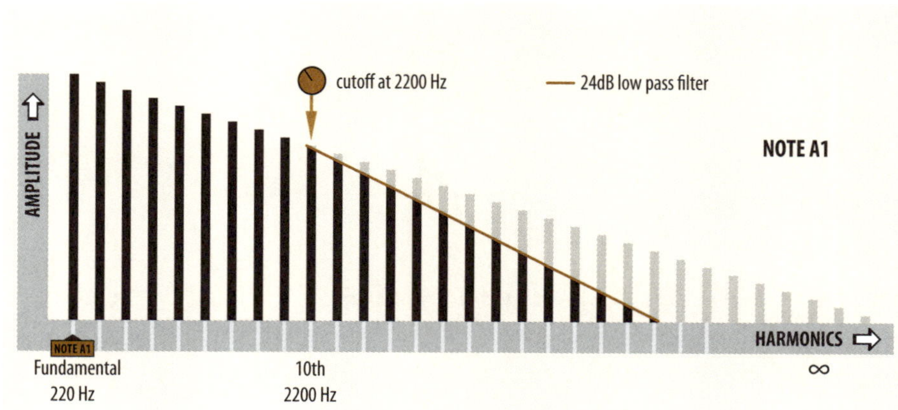


A1-note playing. Correct setting of the cutoff frequentie for a particular sound.

If we take a sound and play it at a low pitch, it sounds good. But if we play it higher, without using the keyboard to control the cutoff frequency, then the sound will start to get muddy. See figure 4.66.

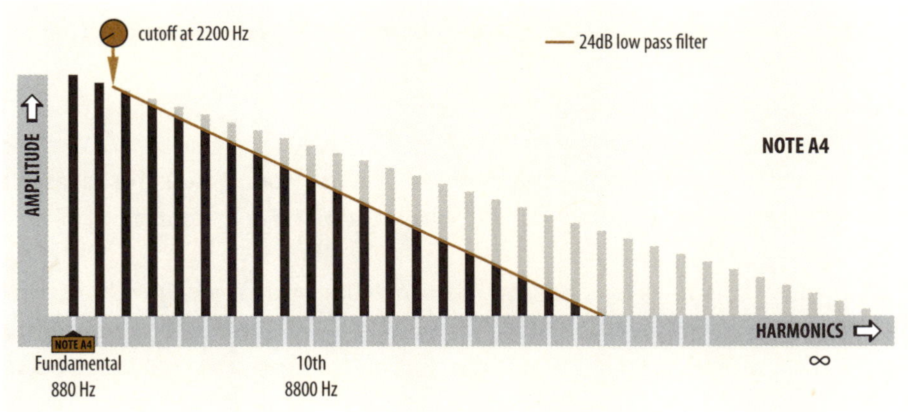


Figure 4.66

A4-note playing. Wrong setting of the cutoff frequentie for a particular sound

If we use the cutoff frequency parameter without the keyboard note modulation, the cutoff frequency setting can be good for low notes, but far too steep for higher pitched notes. By connecting the keyboard note information to the cutoff frequency we can correct this. See figure 4.67

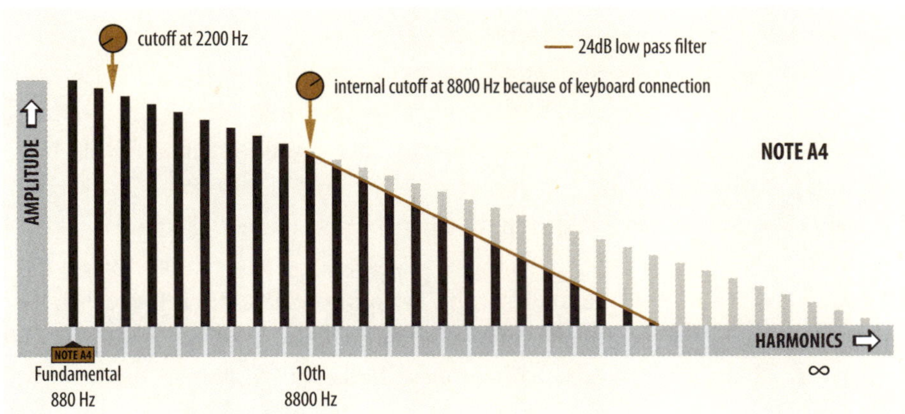


Figure 4.67

Cutoff frequency is now controlled by both the filter setting and the keyboard.

Here's an idea to show this in a different way. This may illustrate why we need keyboard control of the cutoff frequencies.

Imagine a stairway, it represents the keyboard. The higher we are on the stairs, the higher the note. If we go back down to the bottom of the stairs, we can put a hat on.

This hat covers a certain portion of our head.

If we pull the hat down, people cannot see our eyes anymore. Picture this hat as our cutoff frequency and its setting to remove a bit of the upper harmonics.

Well, initially the hat looks good, and cool! The hat is now at a certain horizontal position. It represents the cutoff frequency' setting. If we keep it that way and move up the stairs we'll notice that the hat is not just on our head. It's at our knees now!

This is not good, far from cool and if it reflects the cutoff frequency for sure, it's not okay in sound either. So the hat has to follow us and remain in place as we go up the stairs. If we put it on now, it is again good, and cool! So if the hat stands for our cutoff frequency then it needs to follow the stairs: the pitch. What this tells us is that we need to make our synthesizers and their parameters respond correctly so that we get the sound we expect to get out of them.

Apart from the fact that we expect more harmonics to sound at higher pitches, we simply need to inform the cutoff frequency which note we play and make it respond correctly with the cutoff frequency. The cutoff frequency knob is nothing more than the initial setting which can be okay for a position such as 880Hz, but it will need to follow the pitch to function correctly on the higher notes. So the cutoff frequency needs to move along with the pitch we play to get the sound right.

Of course we can do this manually, but to keep our hands free it is much better to control this by the keyboard note information. Then with the amount knob in between them, we can adjust how much the keyboard influences the cutoff frequency. The more we open the amount knob, the more control the keyboard has in changing the cutoff frequency.

So, to get a more natural feeling this function is added. We humans expect that the higher the pitch is, the sound will be brighter and therefore the cutoff frequency needs to move with the pitch of the note that is played. This is why we'll find keyboard tracking or key follow on almost all synthesizers.

***Element-4: 'Keyboard note information / Note information and Element-2'.***

*Keyboard note information and Element-3*

## Keyboard note information and Element-3

If we look at various synths we're not likely to find a knob or fader that is called key follow or keyboard tracking inside the amplifier section. Using a keyboard would mean that it will control the volume and/or the panning.

Figure 4.68

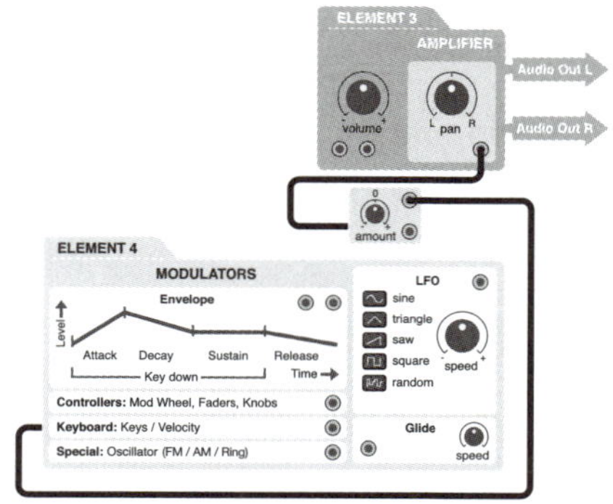


Keyboard controlling panning.

This is not often used but in the case of volume, it can be handy to control the volume in case the sound needs to be corrected at higher or lower pitch ranges. If panning modulation is available, it also can be interesting. Low and high notes will be panned to either side of the stereo image.

WATCH ONLINE

***Element-4: 'Keyboard note information / Note information and Element-3'.***

## Keyboard note information and Element-4

The keyboard note information addressed to Element-4 is mainly used on the envelopes and its times parameters. Controlling an LFO is also now and then used and especially cool for effect sounds.

*Keyboard note information and Element-4*

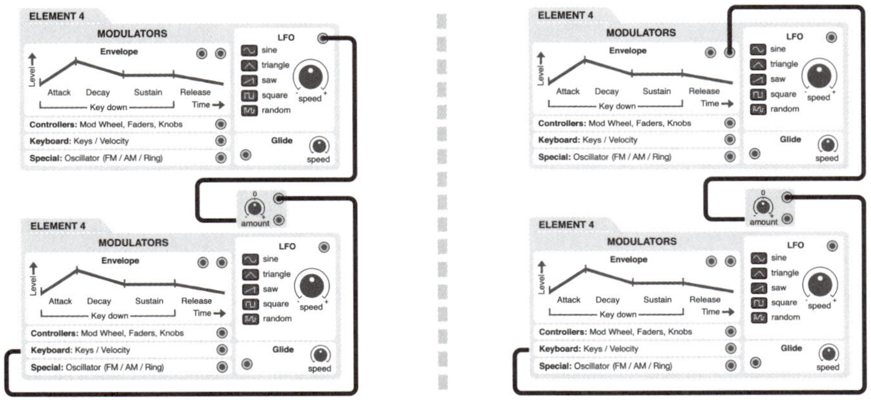


Figure 4.69

Keyboard controlling LFO speed or keyboard controlling envelope time.

***Element-4: 'Keyboard note information / Note information and Element-4'.***

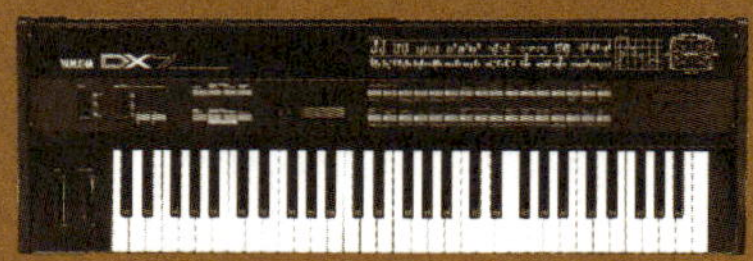

*Keyboard velocity information general*

### Keyboard velocity information general

Keyboard velocity information was not commonly found on the first generation synthesizers. The big revolution on the subject of velocity came with the introduction of the digital Yamaha DX-7 which was in fact a whole new type of synthesis being used in a synthesizer as well.

Not only was velocity and a new synthesis type introduced at that time but just after that came the midi communication interface built into the Roland JX-3p and Prophet-600. Midi caused a revolution and gave us the ability to connect synthesizers together as well as computers. After the introduction of the DX-7, most of the new synthesizers had velocity response, which allows us to play piano (soft) as well as forte (loud) on the keyboard.

The keyboard information we've been explaining up to this point is the note information. Now we also have the information on how hard we play a note. We call this information velocity. In fact, velocity is the correct word for it, because the sensors inside our keyboard measure how fast the key is going down. So if we hit the key hard, the velocity is faster than if we hit the key softly.

This velocity information is most commonly used for volume of course. This is the first thing that we humans expect. Logically, because the softer we play at an acoustic instrument, the softer the volume. Not only the volume changes, but also the timbre changes. Or in other words, the harmonics will change if we play soft or hard. The sound becomes brighter the harder we play. This is also very natural in the acoustic world. The harder we play, the louder the upper harmonics become.

If we look at the 4ES chart, we see that the harmonics are controlled by the filter and the volume is controlled by the amplifier.

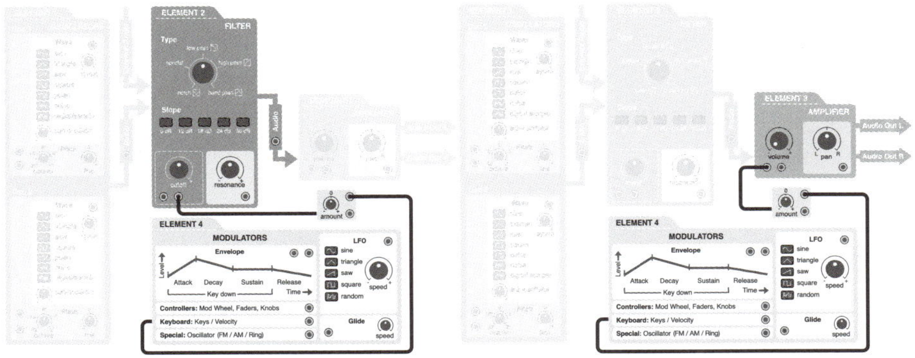


Figure 4.70

So connections to the cutoff frequency and volume are the basic ones we use for keyboard velocity. In between the connection of the keyboard and the parameters we need an amount control, so that we can adjust the responses. The velocity is great for adding expression to a sound. Using velocity with the filter and the volume are not the only options available to us. We can use velocity for controlling other parts of the synthesizer.

WATCH ONLINE ***Element-4: 'Keyboard velocity information / Introduction'.***

*Keyboard velocity information and Element-1*

## Keyboard velocity information and Element-1

Velocity works very well when connected to the pitch or waveform component of an oscillator. To avoid that this connection is always at maximum modulation, we add an amount knob in between the keyboard velocity and the destination so that we can fine-tune our settings. With velocity we have the option to ad minimal changes for making a sound more dynamic while playing. The setting of the amount plays an important role here. If the amount of the velocity control is too deep, then the sound is not changing natural. But if we pick a more minimal amount, it all sounds very dynamic and great.

We get a typical oscillator sync sound when the velocity is modulating the oscillator pitch which is in sync mode. We'll notice there is no pitch change, but far more a timbre change. As explained in the oscillator sync chapter, the oscillator that is set to sync has no life of its own.

Figure 4.71

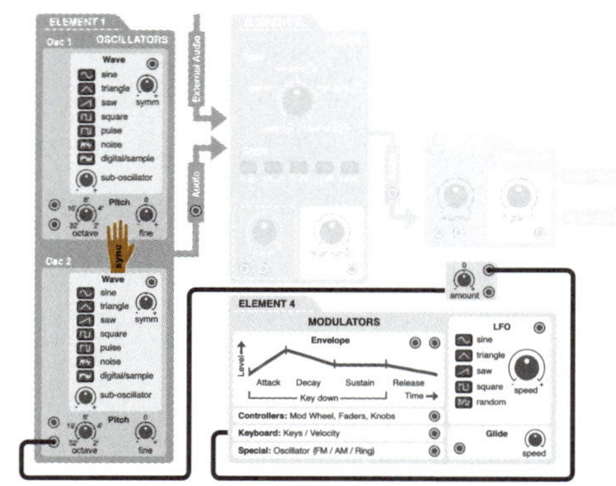


Hier misschien toch een onderschrift?

Changing the pitch of the oscillator now changes its timbre. If we set the oscillator back to its normal mode, or with other words shut down the oscillator sync, we would hear big changes in pitch. Without oscillator sync this connection is not so useful. Still with sync it shows us that velocity controlling the pitch of the oscillator can make sense!

Another useful sound is the pulse waveform modulated by the velocity. By connecting the keyboard velocity to the waveform we change the symmetry of this waveform.

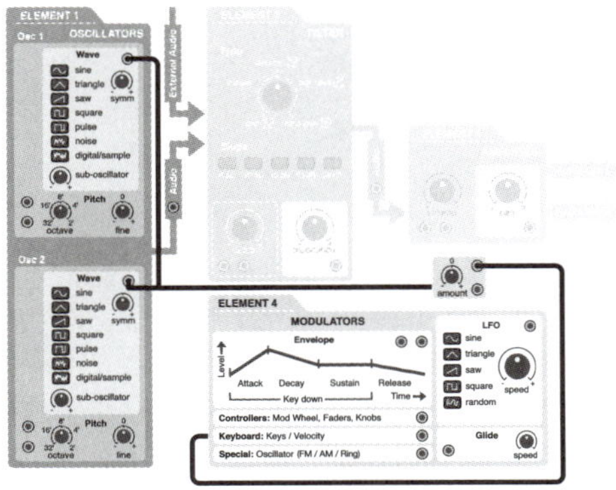

Hier misschien toch een onderschrift?

Figure 4.72

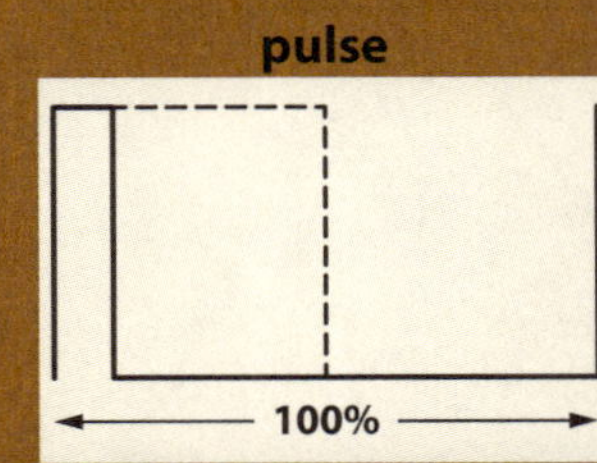


Among the classic waveforms sinus, triangle, saw and square or pulse waveform, the square or pulse waveform is most useful for this. For more detail about symmetry and the pulse or square waveform, please check chapter Element-1.

Todays synthesizers like Predator offer also other type of waveforms, called spectrum waveforms. With these waveforms it can also be very interesting to modulate symmetry using the velocity. Even small changes to the symmetry of a waveform can lead to interesting sounds. There is one thing that we have to keep in mind with this feature. The oscillator waveform symmetry is modulated by velocity. But if we look at the symmetry parameter itself inside the oscillator, then we can see that this can be changed as well. So the symmetry control itself is also the initial setting. The correct setting of the amount control is essential, but also the initial setting of the symmetry parameter itself is important.

WATCH ONLINE ***Element-4: 'Keyboard velocity information / Velocity and Element-1'.***

*Keyboard velocity information and Element-2*

### Keyboard velocity information and Element-2

We already explained that velocity is expected to have results at the volume of a sound and the timbre of a sound. The timbre we usually control inside a subtractive synthesizer with Element-2, the filter. The filter can alter the harmonics that are generated by the oscillators.

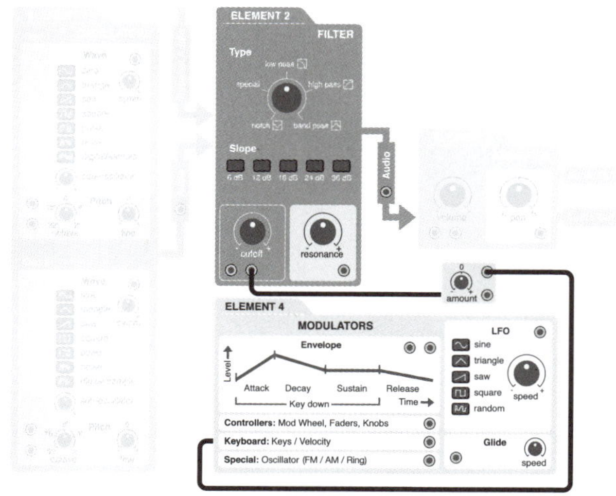


Figure 4.73

Using keyboard velocity makes us control the filter in a very dynamic way. It is the human input that controls the synthesizer part, unlike for instance the LFO or envelope. Looking at the chart of the 4ES reveals us that cutoff frequency and resonance are the only parameters that can be controlled by velocity.

The 4ES also reveals to us that the cutoff frequency has two inputs, compared to the single resonance input. These amounts of inputs are pure symbolic and only tell us that cutoff frequency is used a lot and the resonance input is not used that often. The connection of velocity towards the filter is mostly done using the cutoff frequency parameter. In between we set an amount control, so that we can adjust the impact of the velocity on this cutoff frequency parameter. Of course it depends on the type of sound we seek and how much we like to change the filter cutoff frequency. It can be firm, or very minimal.

With a standard setting, the harder we play the more bright the sound gets. This is what we expect to hear and adds a lot of dynamic feel to the sound. But it also depends on the type of filter we use! With a band pass filter type, the change of the sound is very different. The classic modulation by velocity of the filter is only with the low pass filter types.

***Element-4: 'Keyboard velocity information / Velocity and Element-2'.***

*Keyboard velocity information and Element-3*

## Keyboard velocity information and Element-3

Volume is the first thing that comes to mind if we talk about keyboard velocity. Of course, it depends very much on the personal taste of the player and the sound if we like to have a lot, minimal or maybe no velocity at all at the volume of a sound.

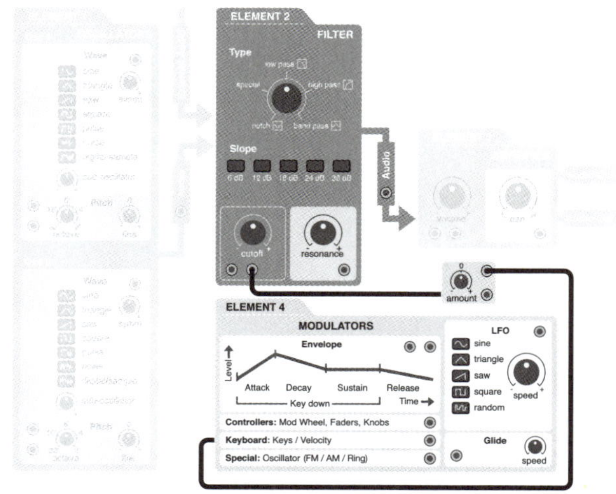


Figure 4.74

It is often overlooked that we can control this velocity response. So instead of editing our track inside a sequencer to have less dynamic changes in volume, it's maybe wise to save a bit of time and control the velocity response at the volume parameter inside our synth.

The amount that controls the velocity modulation of the volume is often hidden inside a modulation matrix or a menu. Still, it's an important parameter because it arranges the dynamic changes of the volume. It depends very much on style of playing or sound expression how much velocity we need to address.

***Element-4: 'Keyboard velocity information / Velocity and Element-3'.***

## Keyboard velocity information and Element-4

*Keyboard velocity information and Element-4*

The keyboard velocity information addressed to Element-4 is mainly used on the envelopes and its time parameters attack, decay, and release. Controlling an LFO is also an option of course and especially cool for effect sounds.

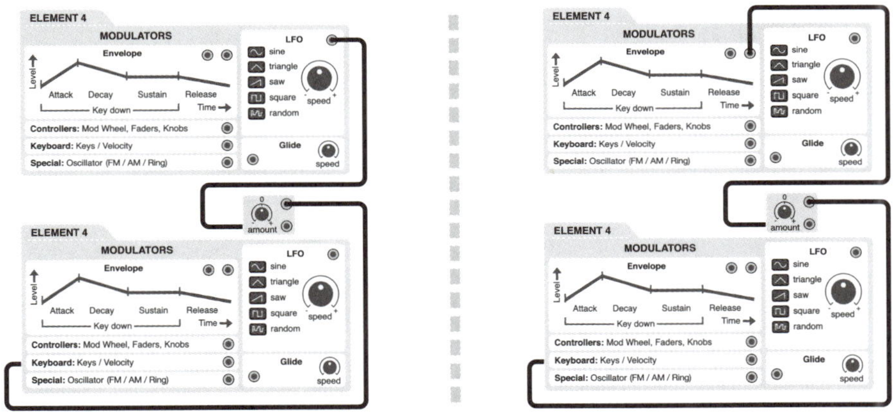


Figure 4.75

Also other parts of Element-4 and even the amount control can be controlled by keyboard velocity. Overall it's used to make dynamic changes in the envelopes while playing.

*Element-4: 'Keyboard velocity information / Velocity and Element-4'.*

*Controllers in general*

## Controllers in general

If we look at the 4ES chart we see controllers are listed. Generally, controllers are physical interfaces used to control parts of the synthesizers. What makes them different from envelopes and LFO's is the fact that the real time human factor counts when using them.

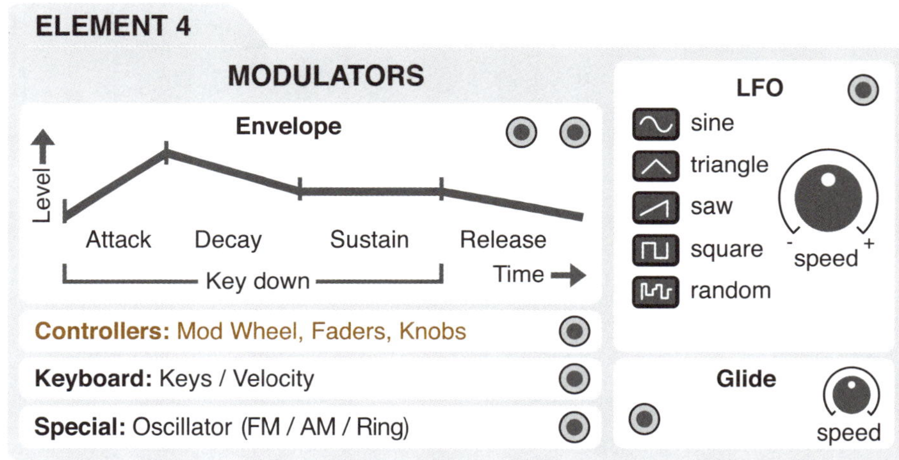


Figure 4.76

Just like when we're playing our keyboard, the human factor is generating the modulation. And so the results of using pitch wheel, modulation wheel or ribbon controller is determined by what we as player do with them.

There are several types of controllers. The most commonly used and available are the pitch wheel, modulation wheel and aftertouch. But there are knobs and faders on synthesizers that can be used as controllers as well. If a synthesizer like for instance an Access Virus b sends out the knob information as a controller, we can control things on other synthesizers with that knob.

***Element-4: 'Controllers in general'.***

- **Pitch wheel controller**

Looking at the 4ES chart tells us that we have to connect the pitch wheel to the oscillator's pitch components.

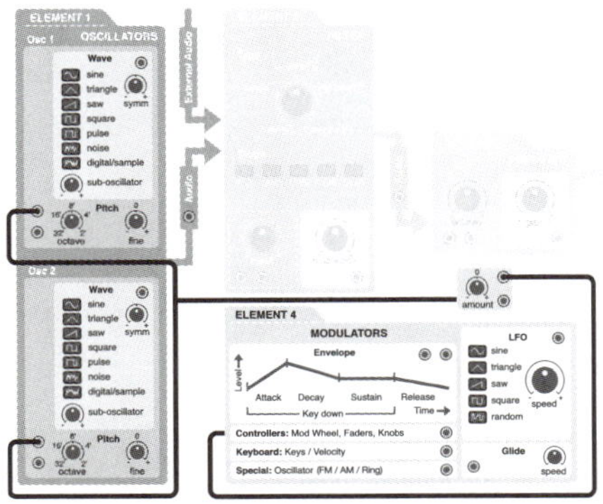


Inside the 4ES chart, we have two oscillators and this means that we'll need to connect to both of them. If we connect only one oscillator then the resulting sound we play is odd when we start using the pitch wheel. Most of the time we don't have to worry about the pitch wheel. Most brands of synthesizers already have the pitch wheel connected to all oscillators as a default setting. They often use a modulation target called main pitch which means that all oscillators available inside the synthesizer are connected to the pitch wheel.

Today's synthesizers allow us to control even the amount for down bending and up bending. This is an interesting feature. With this feature, we can have settings such as 2 semitones up bending and 5 semitones down bending. The pitch wheel is mostly used with the default settings in use. But, if we have the option to control the function, we are free to do so and do other things as well. Maybe we built a sound that doesn't need the pitch wheel function? In that case, we can shut it down and use the pitch wheel for other purposes. We must think out of the box with synthesizers, even if controls like pitch wheel look very default at first glance.

*Element-4: 'Pitch wheel controller.'*

Figure 4.77

- **Modulation wheel controller**

This controller is very flexible in its use. Although from the start, it was very often used for adding vibrato as we can see on the legendary Mini-Moog. Even on today's synthesizers we can find that it's still very often used for adding vibrato to a sound. But if a sound does not need vibrato, the modulation wheel is often used for controlling the cutoff frequency of the filter. The real time human input is very important and makes the modulation wheel exciting to use.

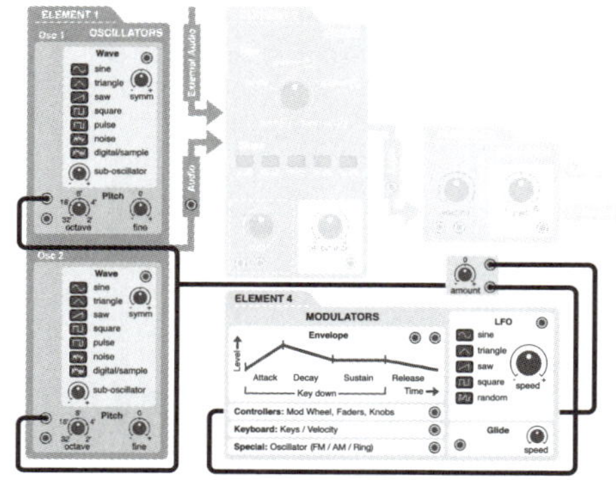


Figure 4.78

Looking at the 4ES chart we can see how to make the modulation wheel connection to control vibrato. An LFO is needed to generate vibrato and we connect this to both oscillators. In between we set an amount control, so that we can set the initial depth of the modulation. With the modulation wheel we need to control the depth and this we accomplish by connecting the modulation wheel to the amount control. So in fact, we can keep the amount control between the LFO and the oscillators set to 0 (zero), because we use the modulation wheel to control the modulation amount. But note, this is really up to us and the sound we're trying to create. We can always have a bit of initial modulation by opening the amount and then add more modulation by using the modulation wheel. Or we can have no initial modulation and only use the modulation wheel to control it. This example is the classic way of using the modulation wheel. But any other parameter can be controlled with the modulation wheel. It's ultimately up to us how we decide to use it.

WATCH ONLINE ***Element-4: 'Modulation wheel controller'.***

- **Aftertouch modulation**

The term aftertouch already tells us how this works. What we need to do as a player is press on the keyboard after playing the notes. A sensor then registers how hard we are pressing and sends out this information as controller information.

Basically, we can compare it in a way to the modulation wheel. The only difference being that we use the pressure on the keyboard to control the amount of modulation. Some keyboards even have polyphonic aftertouch, which means that each note has it's own pressure value. But most keyboards only have monophonic aftertouch. This means if we press a note, the other notes we play will get the same modulation amount.

Take a look at the 4ES on the previous page (figure 4.78) to see the connections inside a synthesizer using aftertouch to control the vibrato of a sound. An LFO is needed to generate vibrato and we connect this to both oscillators. In between the LFO and oscillators we put an amount control, so that we can control the initial depth of the modulation. With the aftertouch we need to control the depth of the modulation and we do this by connecting the aftertouch to the amount control.

So in fact, we can keep the amount control between the LFO and the oscillators set to 0 (zero), and just use the aftertouch to control the modulation amount. We can also have a bit of initial modulation by opening the amount and then add more modulation by using the aftertouch.

This aftertouch controller works great for vibrato, but there are other parameters that can be controlled by it as well. It is up to us to decide how to use it and which parameters we wish to control. So if our keyboard has aftertouch we shouldn't be afraid to use it and experiment.

WATCH ONLINE ***Element-4: 'Aftertouch'.***

- **Modulation knob/fader controllers**

Today's software synthesizers are often controlled by hardware devices. These can be midi keyboards with many knobs on top of them, or a synthesizer like the Virus b that sends out all the knob information as midi controller information.

There are also dedicated midi controllers that focus on editing our software synthesizers like the Novation Controllers. These midi controllers and their knobs and faders make that we have more control and feel over our software synthesizer.

Unlike the modulation wheel, this type of control is not only to dynamically change the sound, but also to build presets and edit sounds during music making. For changing sounds during performance these are for sure also a tool, the same as a regular modulation wheel or aftertouch.

*Element-4: 'Ribbon and other controllers'.*

## Glide / Portamento

*Glide / Portamento*

Inside Element-4 we can also find glide, or portamento as we often call it. This function is mostly used in between the keyboard note output and the pitch input of the oscillators.

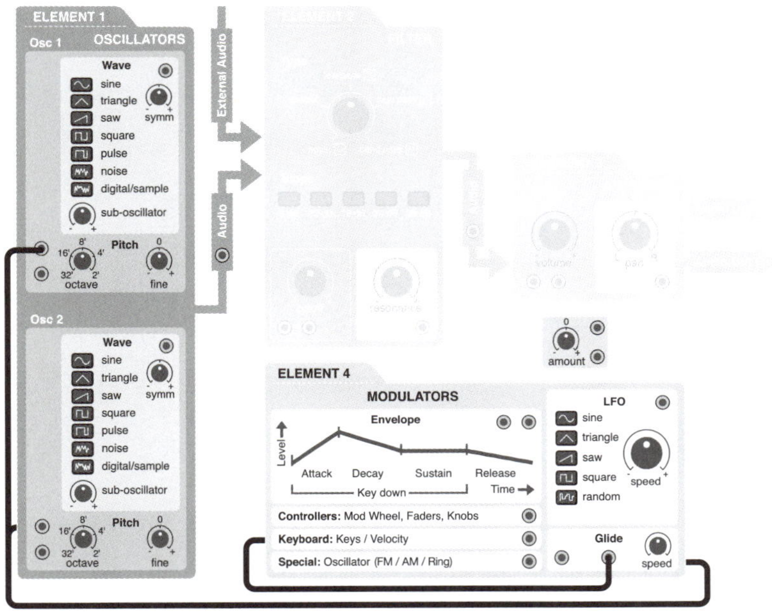


Figure 4.79

Technically glide is a 'lag processor': it limits the rate at which a signal can increase or decrease in value. This means that changes in pitch are not immediate but take a certain time. In analog synthesizers this is done by an electronic circuit and in software this function is simulated by code.

Digital simulations of lag processing - which we for instance can find in Predator - , usually fall into two categories. 'Constant rate', in which the slew from one frequency to another proceeds at a constant speed every interval of time, and 'constant time', in which the slew increases or decreases depending on the frequency interval to be covered, so that the amount of time it takes from one note to the next is always the same.

Predator and other synths also offer the option to activate the glide by playing legato. During normal playing the glide is not active, but if we play legato notes the glide is activated.

The main glide parameter appears in different names on various synthesizers, such as 'speed', 'amount' or 'time'. If the glide has two modes we can also see 'rate' or 'time', depending on the setting selected.
Glide is a cool feature for lead sounds, bass sounds if the legato mode is possible, but also for polyphonic effect sounds.

WATCH ONLINE

***Element-4: 'Glide'.***

**About FM modulation, AM modulation and ring modulation**

*About FM modulation, AM modulation and ring modulation*

This whole training is meant to make synthesizers easier to understand by explaining the various parts and techniques for their use. When it comes to modulation, some synthesizer parts are less often used, such as ring modulation, AM modulation and FM modulation. The musical outcome of these techniques are not always predictable, so therefore we're only going to provide some essential information about these features.

- **FM modulation**

Let's start with FM modulation and wipe away some common confusion about it. FM modulation makes some users think of FM synthesis, which you can find on the DX-7, FM7 and FM8, BLUE and many other synths. FM synthesis is in fact a different type of synthesis, it is actually 'phase modulation'. So instead of modulating the pitch of the oscillators or operators (as they are often called), it modulates the 'phase' of the waveform.

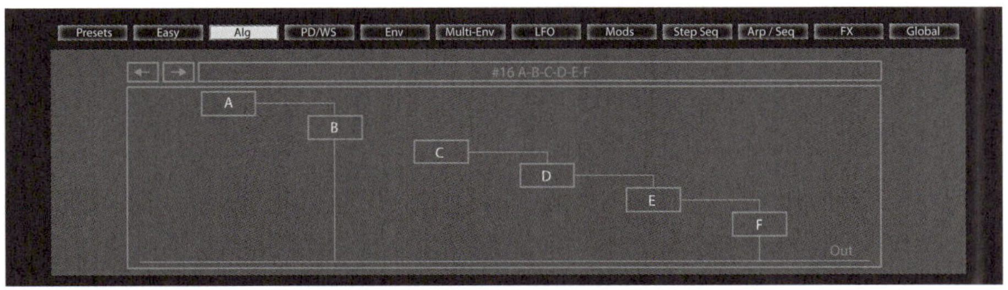

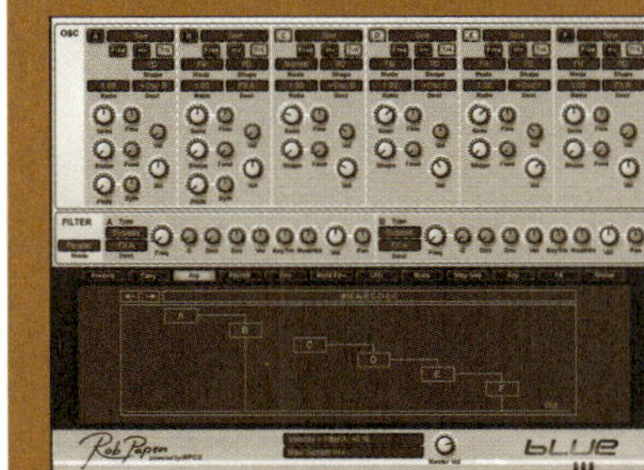

With FM synthesis various oscillators are connected to each other. Some of the oscillators are used to modulate the 'phase' of other oscillators.

An interesting point is that with some subtractive synthesizers, we can also do a bit of FM synthesis type of sounds. It still sounds different, because with real FM synthesis the phase of the waveform is modulated. On the Video there is an example we did with Predator.

With FM modulation inside subtractive synthesis, an oscillator is connected to another oscillator. This is the same connection we'd make if we are going to use an oscillator as an LFO to create vibrato. The difference now is that the oscillator connected plays at a much high (audible) frequency, resulting in an atonal metal sound. Great for bells and effect sounds.

The access Virus b offers this FM (frequency modulation) inside oscillator -2, which means that oscillator-2's frequency is modulated by oscillator-1. So while FM synthesis is an entirely different type of synthesis, FM modulation is nothing more than modulating the frequency. For example, this can be an LFO modulating the pitch-frequency or an envelope modulating the cutoff frequency of a filter. It is just plain old modulation. Of course, the results of modulating and oscillator pitch with another oscillator is very different in sound compared to using an LFO or Envelope. Therefore it appears very often as a separate amount parameter, like we saw on the Access Virus b.

Figure 4.81

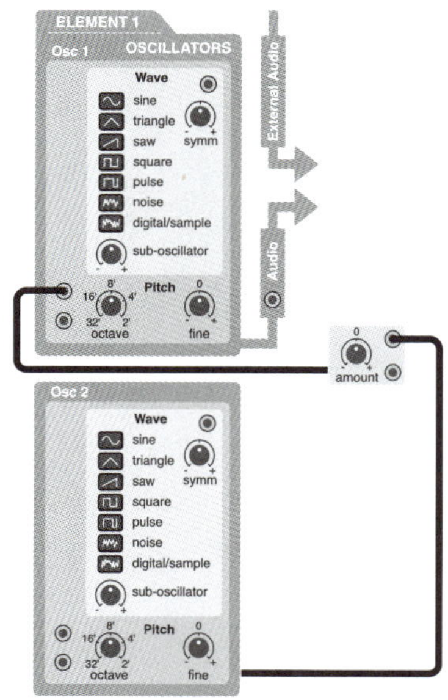


Oscillator 1 frequency modulation using oscillator 2.

***Element-4: 'Special modulation'.***

- **AM modulation**

Looking at the 4ES chart reveals that the output of the oscillators also has a modulation input, so we can modulate its level. This is what we call AM or amplitude modulation, because basically it is the same as if we were using the amplifier volume control to change the volume. However this term tells us that it is used to get different results than if we used modulation done by regular oscillators.

If we were to use an LFO, and connect it to the oscillator output, then the outcome would be a tremolo FX. AM modulation is very close to 'ring modulation'. The difference is that the output of the modulating oscillator is positive only. An oscillator signal has both positive and negative values. With AM modulation, only the positive signal is used.

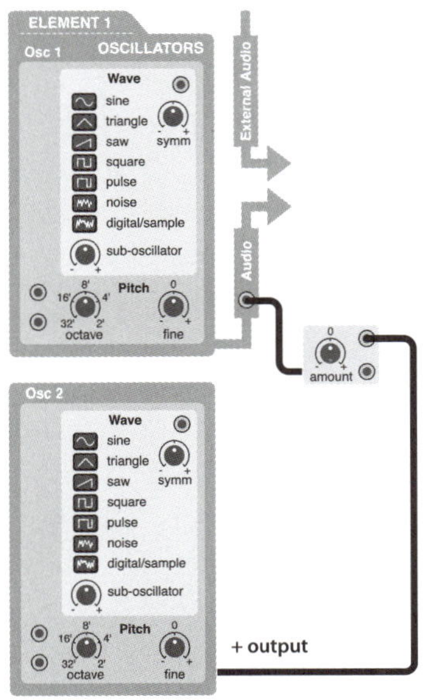


Figure 4.82

Oscillator 1 AM modulation using oscillator 2.

With the 4ES (see figure 4.82) we can also set oscillator-2 to modulate oscillator-1 using an AM setting. Now, oscillator-2 is not connected to the filters anymore but instead modulates oscillator-1.

By using a very low pitch from oscillator-2 we can hear the tremolo FX. The volume output or in other words, the amplitude of oscillator-1 is what is really being modulated. If we keep going higher in pitch the outcome becomes very different. At these very high pitches, we can hear a metallic sound being created.

This metallic sound is great for FX sounds. We can easily build a tuned percussive sound by using an extra envelope to control volume of the modulating oscillator. Once we have found the right setting, the modulation is only at the very beginning of the sound and this will add a very nice metallic overtone to create tuned percussive sounds. Watch the Video for an example.

Just like ring modulation and FM... it's really trial and error with these type of modulations. Results are hard to predict, but if we're lucky, we can sometimes get some great results.

***Element-4: 'Special modulation'.***

- **Ring modulation**

Ring modulation is great for FX sounds and especially for making metallic sounds. What's happening here is that the output of oscillator-1 is multiplied by oscillator-2. The ring amount controls the mixing between ring modulated output and the original output. Ring modulation is very close to 'AM modulation'. The difference is that the output of the modulating oscillator can be positive and negative, an oscillator signal normally has both positive and negative values. With AM modulation, only the positive signal is used.

Figure 4.83

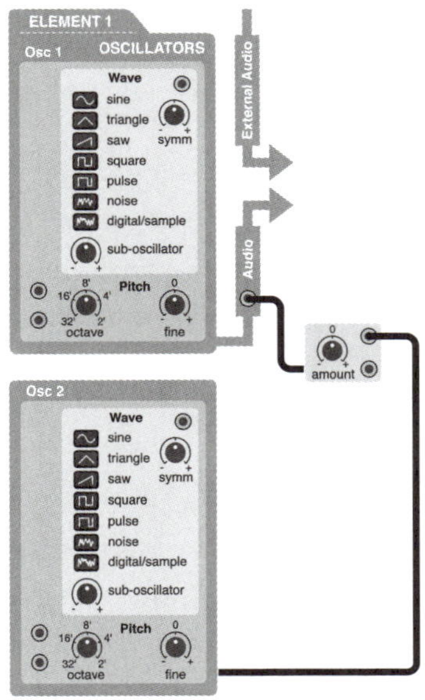


Oscillator 2 ring modulating oscillator 2.

The outcome of this type of ring modulation is far from predictable, but it can be great for FX sounds and metallic sounds. Tip: for best results, change the pitch of both oscillators dramatically and the best waveform to use is a sinus.

***Element-4: 'Special modulation'.***

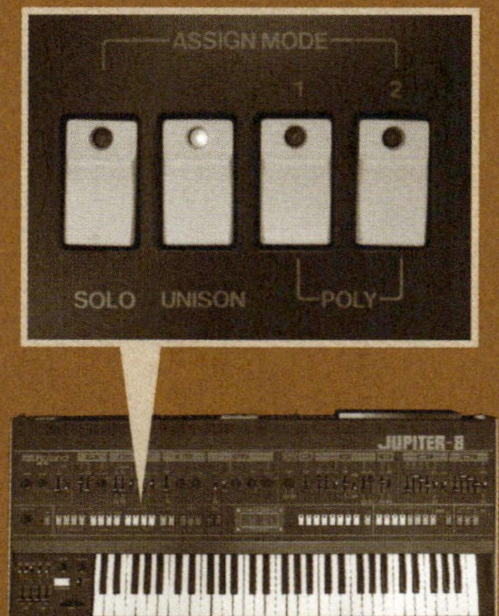


*Unison mode*

## Unison mode

Looking at the 4ES chart, we don't see Unison listed. This mode requires a polyphonic synthesizer and the image below shows clearly why.

Figure 4.84

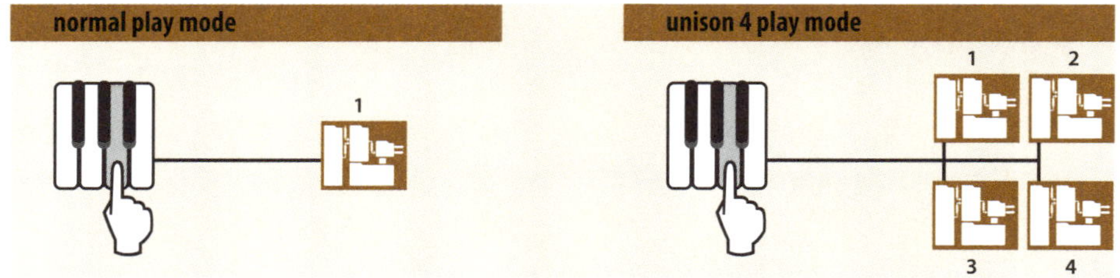


Normal play mode and example with unison 4 play mode.

In general music terms, unison refers to multiple voices or instruments sounding at the same pitch. Singing in unison by a choir, for instance. Many polyphonic synthesizers offer this option as well. Older analog synthesizers, like the Roland Jupiter-8, have the option to play in unison. You can stack its eight voices to one note.

You might expect a massive sound, however that is not always the case! The unison sound is at its best if the 'tune' function of the Jupiter-8 is not used. The tune function brings all the analog oscillators in tune, since even during operation they might get slightly out of tune. In case of unison, when the oscillators are a bit out of tune we have a more massive sound. Hitting the tune button at the Jupiter-8 makes this specific unison quality go. This is the reason why on modern synthesizers like Predator we see the parameter 'unison detune'. If the tuning of the digital oscillators is too perfect we get a phased sound, but if it's slightly imperfect we get a much fuller sound. Keep in mind that using unison on software synthesizers demands more CPU power. This is because you use several voices of the synthesizer for each note you play. In general the use of unison depends on the sound you seek. It's wise to use it with discretion, 'less is often more' applies to unison just as it does to music in general.

***Element-4: 'Unison mode.'***

## Chord memory

*Chord memory*

Chord memory is a feature which can be found on early polyphonic analog synthesizers. Also today's synthesizers often have this feature.

In most cases the chord memory is a temporary memory of the chord you have played. Changing from preset or shutting off the synthesizer will cause this temporary memory to be lost.

Software synthesizers offer a lot more freedom, which we can see inside Predator. Here a chord can be memorized inside the presets. This feature is cool for techno sounds, and also for spacy synthesizer sounds.

***Element-4: 'Chord play modes'.***

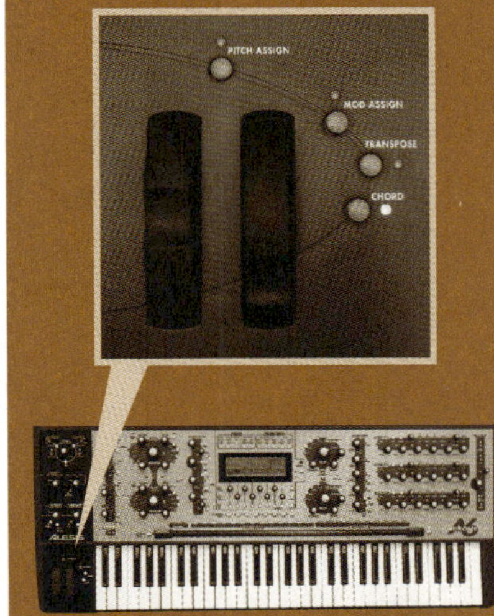

# The Video Menu

| | |
|---|---|
| Video | Introduction > General introduction |
| Video | Introduction > Brief introduction subtractive synthesis |
| Video | Introduction > Intoduction 4 Element Synth |
| Video | Introduction > Before you move on |
| Video | Element-1 > Introduction Element-1 |
| Video | Element-1 > Waveforms part 1 |
| Video | Element-1 > Harmonics |
| Video | Element-1 > Waveforms part 2 |
| Video | Element-1 > Waveform examples > Sinus |
| Video | Element-1 > Waveform examples > Triangle |
| Video | Element-1 > Waveform examples > Saw |
| Video | Element-1 > Waveform examples > Square |
| Video | Element-1 > Waveform examples > Pulse |
| Video | Element-1 > Waveform examples > Symmetry modulation |
| Video | Element-1 > Waveform examples > Noise |
| Video | Element-1 > Waveform examples > Digital |
| Video | Element-1 > Waveform examples > Samples |
| Video | Element-1 > Sub-oscillator |
| Video | Element-1 > Waveform modulation |
| Video | Element-1 > Oscillator pitch > In general |
| Video | Element-1 > Oscillator pitch > Octaves detuning |
| Video | Element-1 > Oscillator pitch > Fine detuning |
| Video | Element-1 > Oscillator pitch > More than two oscillators |
| Video | Element-1 > Oscillator pitch modulation |
| Video | Element-1 > Oscillator sync |
| Video | Element-1 > Oscillators in general > More is better? |
| Video | Element-1 > Oscillators in general > Start phase |
| Video | Element-1 > Oscillators in general > Bass sounds |
| Video | Element-1 > Conclusion |

WATCH ONLINE

| | |
|---|---|
| Video | Element-2 > Introduction Element-2 part 1 |
| Video | Element-2 > Introduction Element-2 part 2 |
| Video | Element-2 > Filter types |
| Video | Element-2 > Cutoff frequency |
| Video | Element-2 > Resonance |
| Video | Element-2 > Resonance uses |
| Video | Element-2 > Low-pass filter |
| Video | Element-2 > High-pass filter |
| Video | Element-2 > Corrective high-pass filter |
| Video | Element-2 > Band-pass filter |
| Video | Element-2 > Notch filter |
| Video | Element-2 > Special filter |
| Video | Element-2 > Cutoff frequency modulation |
| Video | Element-2 > Using envelope modulation |
| Video | Element-2 > Cutoff frequency keyboard note modulation |
| Video | Element-2 > Using LFO Modulation |
| Video | Element-2 > Special cutoff modulation |
| Video | Element-2 > Resonance modulation |
| Video | Element-2 > Conclusion |
| Video | Element-3 > Introduction Element-3 |
| Video | Element-3 > Volume parameter |
| Video | Element-3 > Three types of volume |
| Video | Element-3 > Amplifier envelope |
| Video | Element-3 > Exceptions |
| Video | Element-3 > Panning |
| Video | Element-3 > Conclusion |

Video Element-4 > Introduction Element-4
Video Element-4 > Are all movements arranged by Element-4?
Video Element-4 > Modulation amount
Video Element-4 > Envelopes > Introduction
Video Element-4 > Envelopes > How an envelope works
Video Element-4 > Envelopes > Attack
Video Element-4 > Envelopes > Decay
Video Element-4 > Envelopes > Sustain
Video Element-4 > Envelopes > Release
Video Element-4 > Envelopes > Working with envelopes
Video Element-4 > Envelopes > Envelope and cutoff frequency
Video Element-4 > Envelopes > Positive and negative modulation
Video Element-4 > Envelopes > Sustain fade
Video Element-4 > Envelope and Element-1 pitch
Video Element-4 > Envelope and Element-1 waveform
Video Element-4 > Envelope and Element-2
Video Element-4 > Envelope and Element-3
Video Element-4 > Envelope and Element-4
Video Element-4 > Trigger/play modes
Video Element-4 > LFO > Introduction
Video Element-4 > LFO > LFO speed
Video Element-4 > LFO > LFO speed modulation
Video Element-4 > LFO > LFO waveforms
Video Element-4 > LFO > Sinus, saw and pulse
Video Element-4 > LFO > Random and sample&hold
Video Element-4 > LFO > Waveform modes
Video Element-4 > LFO and Element-1 > Introduction
Video Element-4 > LFO and Element-1 ...... Pitch modulation FX sound
Video Element-4 > LFO and Element-1 ...... Pitch modulation tip 1
Video Element-4 > LFO and Element-1 ...... Pitch modulation tip 2
Video Element-4 > LFO and Element-1 ...... Waveform modulation
Video Element-4 > LFO and Element-2
Video Element-4 > LFO and Element-3
Video Element-4 > LFO and Element-4

WATCH ONLINE

Video Element-4 > Keyboard introduction

Video Element-4 > Keyboard note information > The basic connection

Video Element-4 > Keyboard note information > Wave modulation

Video Element-4 > Keyboard note information > Note information and Element-2

Video Element-4 > Keyboard note information > Note information and Element-3

Video Element-4 > Keyboard note information > Note information and Element-4

Video Element-4 > Keyboard velocity information > Introduction

Video Element-4 > Keyboard velocity information > Velocity and Element-1

Video Element-4 > Keyboard velocity information > Velocity and Element-2

Video Element-4 > Keyboard velocity information > Velocity and Element-3

Video Element-4 > Keyboard velocity information > Velocity and Element-4

Video Element-4 > Sequencers and arpeggiators

Video Element-4 > Controllers in general

Video Element-4 > Pitch wheel controller

Video Element-4 > Modulation wheel controller

Video Element-4 > Aftertouch

Video Element-4 > Ribbon and other controllers

Video Element-4 > Glide

Video Element-4 > Special modulation

Video Element-4 > Unison mode

Video Element-4 > Chord play modes

Video Element-4 > Conclusion

## Colophon


| | |
|---|---|
| Concept and idea 'The 4 Element Synth' teaching: | Rob Papen |
| Production supervisor: | Jeroen van Iterson www.musictech.nl |
| Text checkup script and booklet: | Ken Fennel |
| Additional voice over and script checkup: | Laura Escudé www.evotechaudio.com |
| Video editing, authorizing and live recordings: | Jochem Tames www.tamesmedia.nl |
| The 4ES synth image: | Shaun Ellwood www.decoderdesign.com |
| Other image inside book: | Frans Rutten www.imagedealers.nl |
| Book design and build up: | Frans Rutten www.imagedealers.nl |
| Cover picture Rob Papen: | Ivan Willems |
| Additional video recordings: | Mike Papen |

Very special thanks to my wife Marion for her endless support and Jeroen van Iterson for his patience and endless support of this crazy project.

Special thanks also to: Michel van Osenbruggen www.synth.nl for allowing us to shoot in his (synthesizer heaven) studio. Frans Rutten for his outstanding support and feedback in building the book and its layout. Jon Ayres, for helping out and being a great partner in building all the RP synthesizers and plug-ins. Matt Friedman of www.vintagesynth.com for allowing us to use some images for the book and videos. Matthias Becker and Dieter Stork, for allowing me to use the Moog System-55 photo of their own legendary book 'Synthesizer von Gestern'.

Special thanks concerning the reprint of this book: Jason Barratt-Hill and John Small for pointing out some spell check errors in the first print of this book.

Camtasia studio for creating 'the screen-capture software and editing tool'.

Thanks to all the synthesizer enthusiasts who attended my live masterclasses, which encouraged me to perfect the method I use for this masterclass over and over.

Supporting brands in alphabetic order: Access, BigFish audio, Camel Audio, E-mu, Loopmasters, Mac Logic, Native Instruments, Novation, Rob Papen, Roland, Steinberg, Synthesizer.com, Yamaha.

The identifier 'Rob Papen' refers to the Rob Papen Soundware company.
General inquiries should be directed to our home page www.robpapen.com

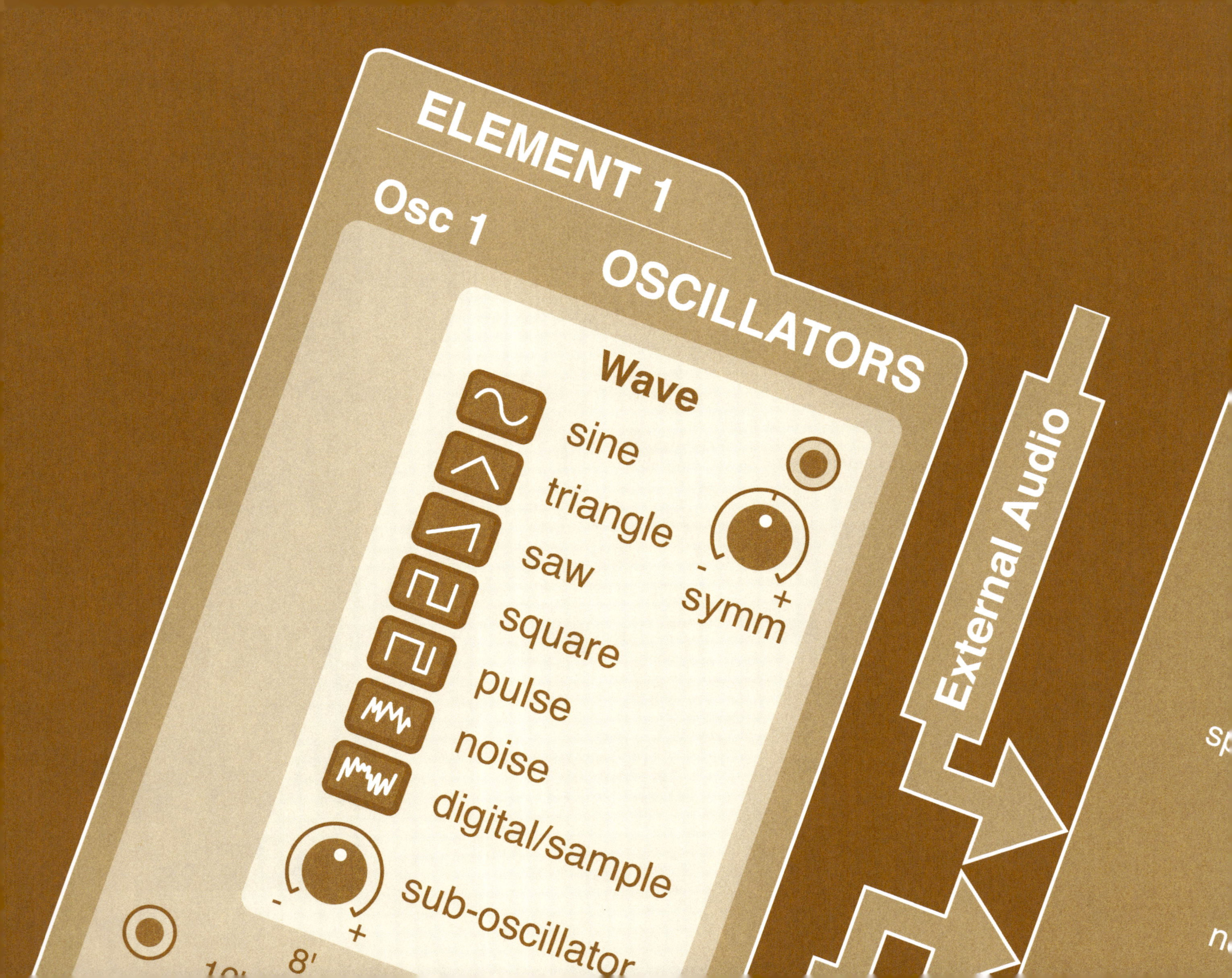
ELEMENT 1
Osc 1
OSCILLATORS
Wave
sine
triangle
saw
square
pulse
noise
digital/sample
-
+
symm
-
+
sub-oscillator
8'
External Audio

## Life is like an oscillator...

Stick to your own wave and you will sound best.

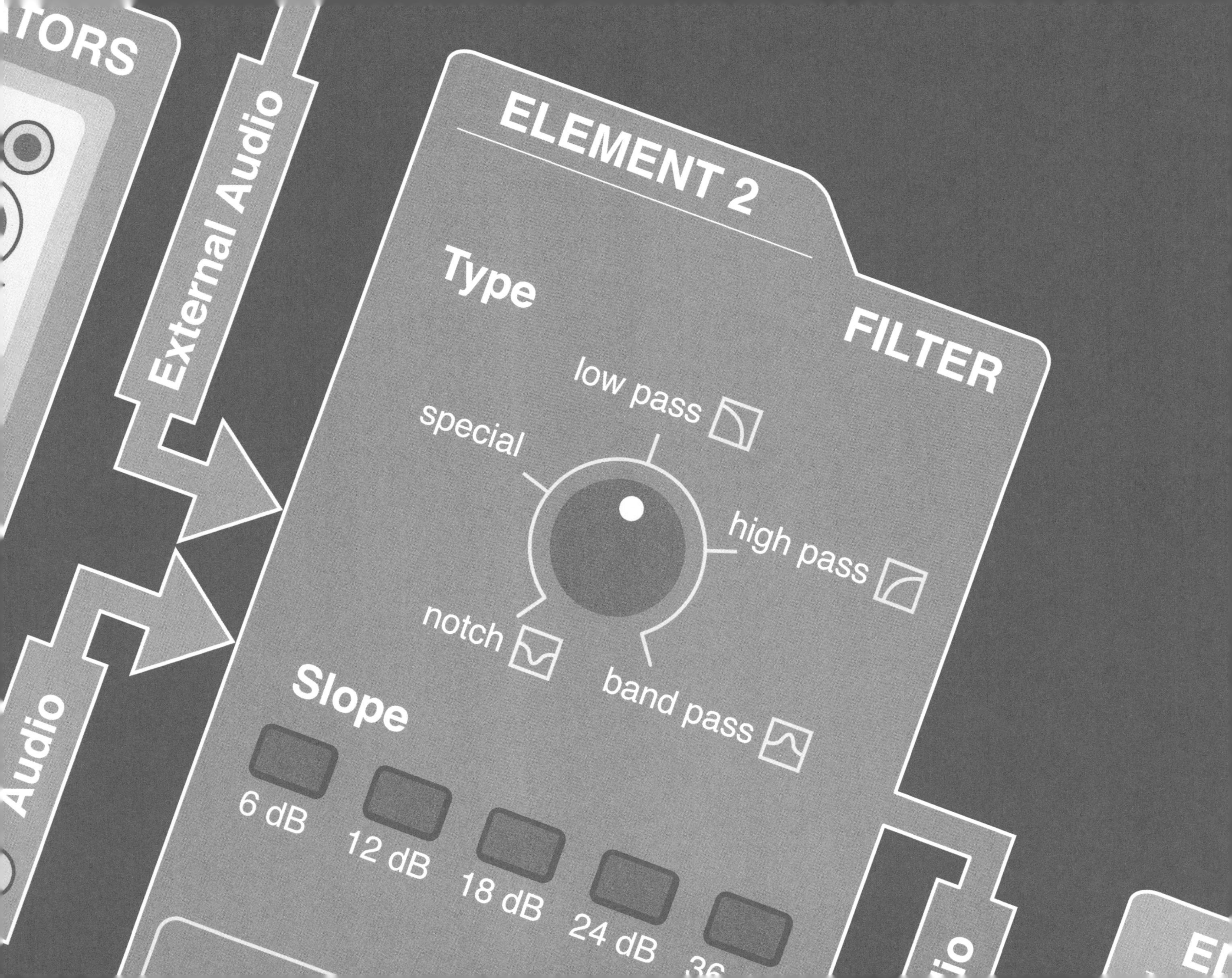
External Audio
ELEMENT 2
Type
FILTER
low pass
special
high pass
notch
band pass
Slope
6 dB
12 dB
18 dB
24 dB
Audio

# Life is like the cutoff of a filter...

Be at the right spot to make a difference.

Audio
ELEMENT 3
AMPLIFIER
-
volume
+
L
pan
R
Audio Out L
Audio Out R
0
-
+
amount

## Life is like an amplifier envelope...

Timing is crucial.

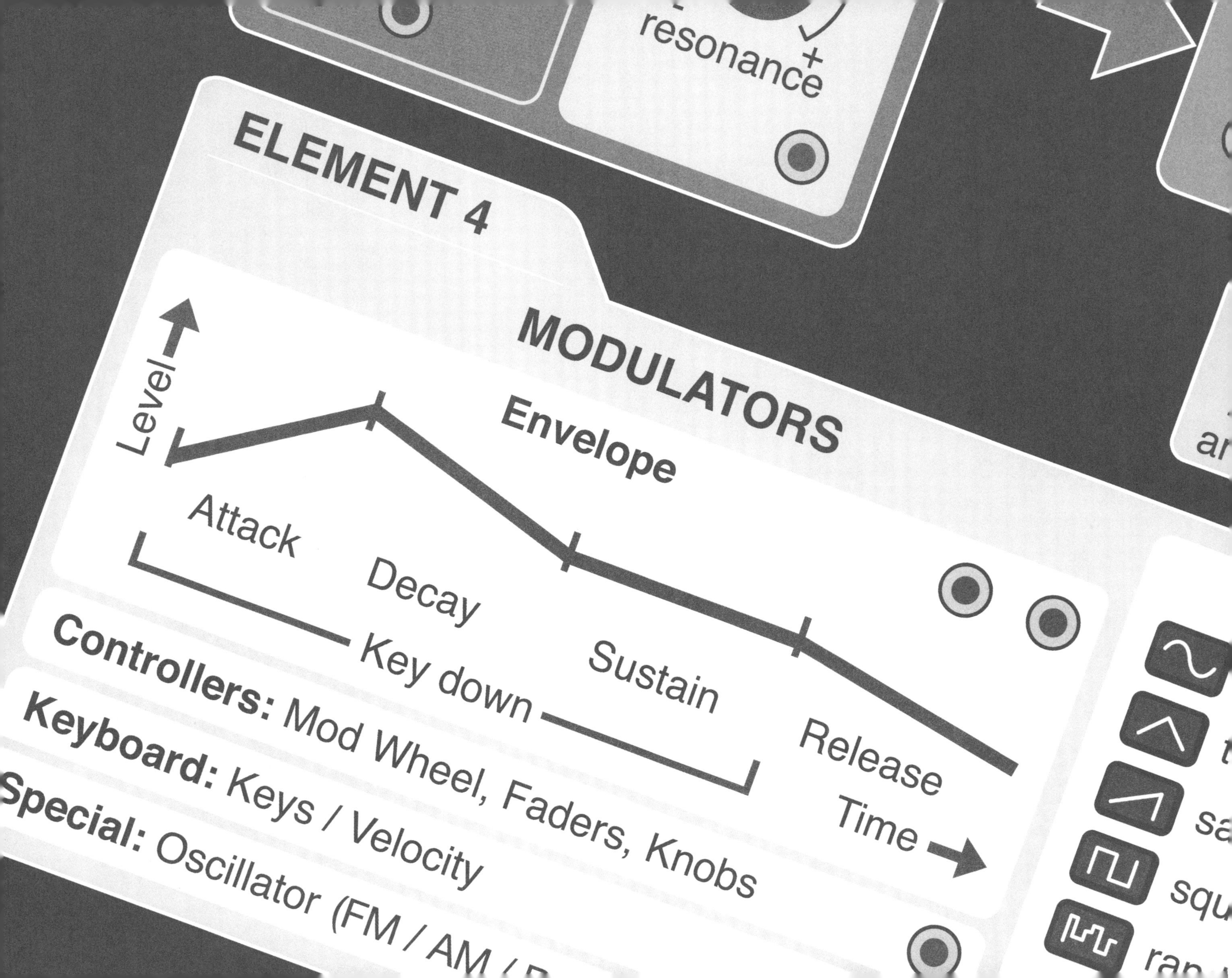

resonance
+
ELEMENT 4
MODULATORS
Envelope
Level
Attack
Decay
Sustain
Release
Key down
Time
Controllers: Mod Wheel, Faders, Knobs
Keyboard: Keys / Velocity
Special: Oscillator (FM / AM /

## Life is like modulation...

You need to address your destination with the right amount.

# THE 4 ELEMENT SYNTH

by Rob Papen

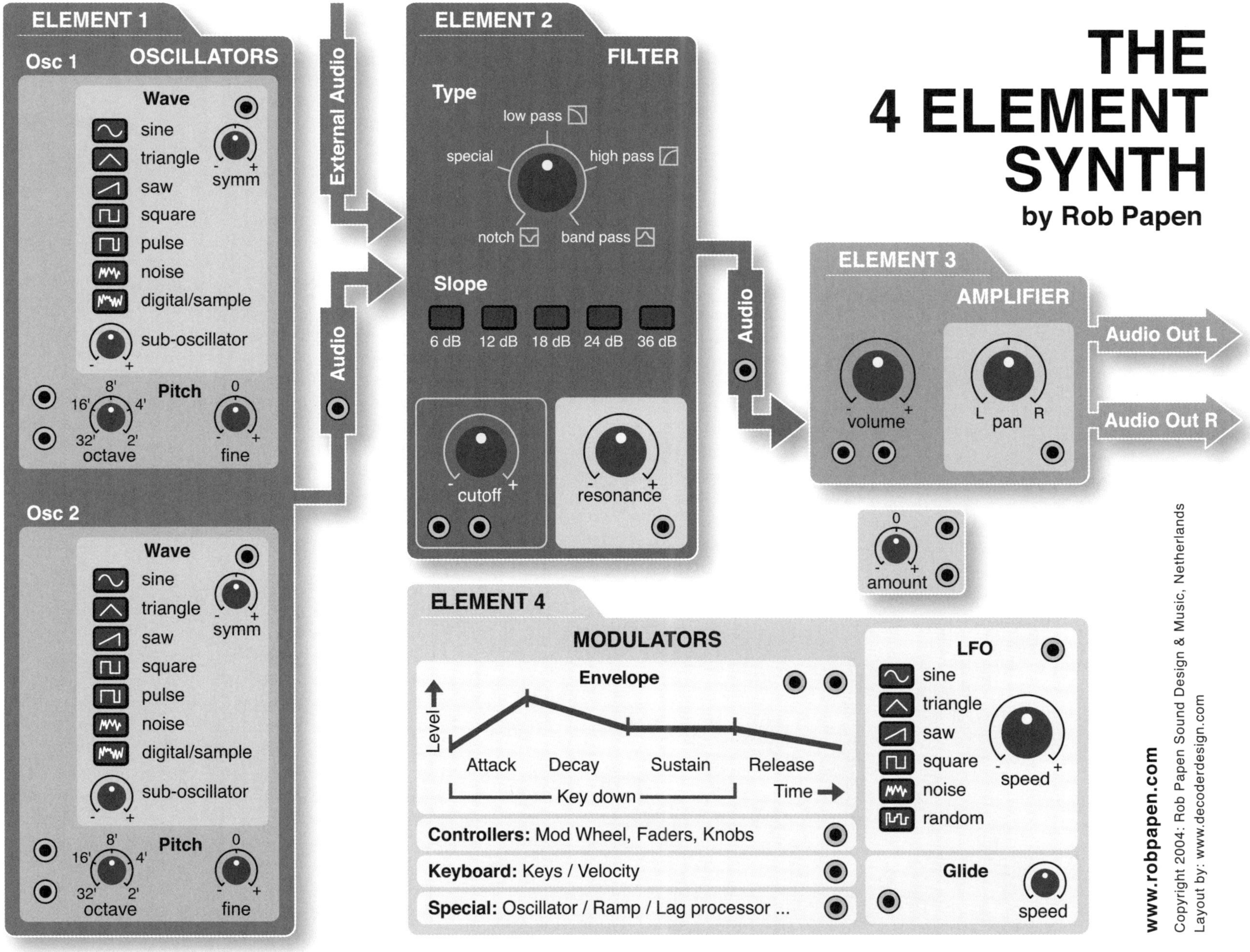


**www.robpapen.com**



Layout by: www.decoderdesign.com